Sergei Vinogradskii and the Cycle of Life

Archimedes

NEW STUDIES IN THE HISTORY AND PHILOSOPHY OF SCIENCE AND TECHNOLOGY

VOLUME 34

Archimedes has three fundamental goals; to further the integration of the histories of science and technology with one another: to investigate the technical, social and practical histories of specific developments in science and technology; and finally, where possible and desirable, to bring the histories of science and technology into closer contact with the philosophy of science. To these ends, each volume will have its own theme and title and will be planned by one or more members of the Advisory Board in consultation with the editor. Although the volumes have specific themes, the series itself will not be limited to one or even to a few particular areas. Its subjects include any of the sciences, ranging from biology through physics, all aspects of technology, broadly construed, as well as historically-engaged philosophy of science or technology. Taken as a whole, *Archimedes* will be of interest to historians, philosophers, and scientists, as well as to those in business and industry who seek to understand how science and industry have come to be so strongly linked.

For further volumes:
http://www.springer.com/series/5644

Lloyd Ackert

Sergei Vinogradskii and the Cycle of Life

From the Thermodynamics of Life to Ecological Microbiology, 1850–1950

Lloyd Ackert
Department of History and Politics
Drexel University
Philadelphia, USA

ISSN 1385-0180
ISBN 978-94-017-8101-5 ISBN 978-94-007-5198-9 (eBook)
DOI 10.1007/978-94-007-5198-9
Springer Dordrecht Heidelberg New York London

Softcover reprint of the hardcover 1st edition 2013

Printed on acid-free paper

Springer is part of Springer Science+Business Media (www.springer.com)

I dedicate this work to my wife Margot Curran and daughter Merle Louise Curran-Ackert.

Acknowledgements

It is a great honor to be considered a student of Daniel P. Todes. He inspired me in selecting Vinogradskii as a topic for my doctoral thesis, in the approach I adopted in exploring it, and encouraging me through its many versions. Through his writings, his wise and frank advice, and countless exchanges of manuscript drafts he has and continues to provide a model of mentorship that I strive to emulate.

My many conversations with Sharon Kingsland contributed significantly to my understanding of the history of biology, evolutionary theory, and ecology. To her I owe the kind and steady impetus ultimately to publish the revised version of the monograph.

My colleague, James Strick, greatly encouraged my work with invitations to present my research to his students at Franklin & Marshal College, and by introducing me to a rich community of soil scientists.

My research for this monograph was supported by an Advanced Graduate Training Fellowship from the Social Sciences Research Council, Eurasia Program. I greatly appreciate the assistance rendered by the staff at the libraries and archives at which I conducted my research: the Archiv Russkoi Akademii Nauk, Marina Sorokina and T. M. Koroleva at its Moscow filial and N. S. Pokhorenko at its St. Petersburg filial; the Biblioteka Akademii Nauk, St. Petersburg; Stephane Kraxner and Anne Weyer of the Service des Archives of the Institut Pasteur, Paris; the manuscript collection and the European reading room of the Library of Congress in Washington, DC; Thomas Frusciano at the Rutgers University Library, New Brunswick; Tom Izbicki at the Milton S. Eisenhower Library and Christine Ruggere at the Welch Library of the Johns Hopkins University, Baltimore; and N. A. Chekmareva at the Tsentral'nyi Gosudarstvennyi Istoricheskii Archiv in St. Petersburg, Russia. I had especially rewarding discussions with Iu. P. Golikov, T. B. Andriushkevich, and Iu A. Mazing at the Museum of the Institute of Experimental Medicine in St. Petersburg, Russia.

At Johns Hopkins University, Gert Brieger, Jeffrey Brooks, Jerome Bylebyl, Pam Long, Harry Marks, and Lawrence Principe guided me along during the course of this project. In Russia, Danil Alexandrovich, Yasha Gall, Eduard Kolchinskii, Mikhail Konashev, and Sergei Orlov improved my knowledge of Russian science. I explored Vinogradskii's place in Russian science and culture with my good friend Evgeny

Khevets (d. 2009). In France, Bernard Vedry, who discovered the lost Vinogradskii tomb, and his wife Brigitte showed great generosity to my family. I owe much of my understanding of Vinogradskii's life in Paris and Brie-Compte-Robert to them. I learned much from my fellow graduate students at Hopkins: Alexa Green, Scott Knowles, Joshua Levens, and David Munns, Keith Barbera, Jesse Bump, Greg Downey, Sander Gliboff, Tom Lassman, Buhm Soon Park, and Karen Stupski—and of the new generations: Allison Kavey, Allison Marsh, Nancy Medley, Maria Portuondo, and Jimmy Shaffer.

Contents

Introduction

Everything that the plants take from the air they give to animals, the animals return it to the air; this is the eternal circle in which life revolves but where matter only changes place.

Jean Baptiste Dumas, 1842[1]

In 1949, at the age of 93, Sergei Nikolaevich Vinogradskii made one final effort to establish his legacy in the history of science. He concluded his scientific career by synthesizing his life's work in a 900-page compendium entitled (in French) *Soil Microbiology: Problems and Methods, Fifty Years of Investigations*.[2] That this work appeared just after World War II, during a time of few resources, and that it was soon after translated into Polish and Russian, demonstrates the high regard for Vinogradskii in the international scientific community. The people and organization—Selman Waksman, Nobel Prize laureate (United States), A. A. Imshenetskii, Soviet Academy of Sciences (Soviet Union), and Pasteur Institute (Paris, France)—that published these volumes reflects the breadth of Vinogradskii's network.

Vinogradskii entitled his book *Soil Microbiology*, but, revealingly, he structured in order to highlight his unique contributions to the history of ecology. Arranging it thematically by research subject rather than chronologically, it is clear he considered the volume to be more than a celebratory monument. He meant it as a reformatory call to action—at once reminding readers of his own legacy and directing them to the progressive ecological significance of his work. As a final statement of his own debt to the past—and perhaps according to the propriety of his French audience—he ended his tome with an essay on "The Principles of Ecological

[1] Jean-Baptiste Dumas, *Essai de Statique Chimique des Etres Organisés*, leçon le 20 août 1841, 2nd édition, (Paris: Fortin, Masson, and Ce, Librairies, 1842).

[2] Serge Winogradsky, *Microbiologie du Sol: Problèmes et Méthodes, Cinquante ans de Recherches* (Paris: Masson et Cie Editeurs, 1949).

Microbiology, A Synthesis." Writing in 1945, he traced "the remote origin of this new branch of the grand microbiological science" to Louis Pasteur's concept of "the role of the '*infiniment petits*' in nature."[3] Here Vinogradskii understood what historians of science have only begun to understand—that ecology owes a substantial debt to microbiology.[4]

Vinogradskii's emergence as an ecologist was hard-won—an analysis of his publications—in their original form—reveals his transformation was gradual: from plant physiologist, to microbiologist, with occasional tangential forays into medical bacteriology, and to ecological microbiologist.[5] At the surface, his story conforms to the contours of history of ecology, usually portrayed as created by botanists who combined Humboldtian phytogeography with Darwinian evolutionary theory during the late nineteenth and early twentieth century. The details of Vinogradskii's life and career, as I explore them in this scientific biography, however, show an alternate story where botany and plant physiology are replaced by microbiology, and phytogeography by the 'thermodynamics of life.' By recognizing these distinctions, a more comprehensive history of ecology emerges—one that emphasizes the role not of natural historians, Darwinists, and plant communities; but rather of experimentalists (who often fused their laboratory investigations with field observations), holists, and soil microbes. In addition, tracing the history of Vinogradskii's activities in some of Europe's foremost research institutions allow us to include a group of scientists and scientific disciplines usually neglected by historians of ecology. The achievements of these microbiologists, plant physiologists, soil scientists, and geobotanists contributed to the formation of ecology through their experimental investigations of energy, matter, and life.

This seemingly disparate array of scientists shared an essential component of their theoretical perspective, one that has only rarely been noticed by historians. Any mention of the cycle of life was, in fact, completely absent from Vinogradskii's *Soil Microbiology*. And yet, it is primarily through exploring his devotion to the cycle of life—and especially his translation of it into an experimental program—that we encounter a new dimension in the history of ecology. Through this concept, we find combined in his research Louis Pasteur's microbiology, Ferdinand Cohn's bacterial taxonomy, and the holistic thermodynamics of Andrei Famintsyn's plant physiology.

Of the histories running in parallel here, central are two: Vinogradskii's personal trajectory and the transformation of the cycle of life from a popular, quasi religious holistic view of nature into an experimental method applied in plant physiology, then microbiology, soil science, and ultimately in ecological microbiology. Thus, we can

---

[3] Serge Winogradsky, "Principes de la microbiologie oecologique: une synthèse, 1945" in *Microbiologie du Sol: Problèmes et Méthodes, Cinquante ans de Recherches*, 839–848; see 839 for quote.

[4] This was recognized at the time of its appearance; See Roger Y. Stanier, "The Life-Work of a Founder of Bacteriology," a review of *Microbiologie du Sol* in *The Quarterly Review of Biology*, Vol. 26, No. 1, Mar., 1951, 35–37.

[5] By original, I mean the initial publications in German, Russian, and French, rather than his edited versions that appear in his 1949 translated/edited form.

follow Vinogradskii's career as he transferred his cycle of life tradition step-wise within a diverse set of scientific schools. The case studies or episodes I present indicate that as novel as Vinogradskii's innovations were, they also reflect the ease by which numerous microbiologists, soil scientists, forestry specialists, and medical bacteriologists integrated them into their own research programs.

Vinogradskii intended his *Soil Microbiology* to be autobiographical. As an historical source, however, it provides a clearer window on his activities and interests in the mid-twentieth century than on the previous decades. He deserves this new biography, not only because he made significant contributions to the sciences of plant physiology, microbiology, soil science, and ecology, but also because his work unites these fields in a novel way. In this exercise in scientific biography, I agree with Thomas Hankins that biographies can provide a window onto the broader understanding of science in its social and cultural context. I treat Vinogradskii's life as the intersection of this historical context and his "scientific, philosophical, social and political ideas."[6] My focus here is on Vinogradskii's scientific research and how it changed and stayed the same over his career. Following the model provided by Thomas Soderquist, Nathaniel Comfort, Frederick Holmes, and other scientific biographers, I use Vinogradskii's 'life and times' to reconstruct significant episodes of his laboratory practices and the role of theory in their development.[7]

The context for much of Vinogradskii's story is Russian. His formative years as a scientist occurred during the period when Russia was developing its own native scientific journals, institutions, and schools. During the 1880s, the Russian government expanded its support for science and employed increasing numbers of Russian scientists in the place of foreign, often Western European, scientists. This is not to say that the strong link between Russian and Western science and medicine was weakened—it may be that during this process it was indeed strengthened. When the government founded new institutes, it looked first to Western European models—for example, Prince Oldenburgskii chose to model his Imperial Institute of Experimental Medicine (where Vinogradskii worked from 1891 to 1912) on the Parisian Pasteur Institute. In addition, scientific journals during this period were often published in both Russian and German or French. While these issues are peripheral to my central aims, we will see that Vinogradskii's story reflects the social and institutional changes that Russian science underwent at the turn of the nineteenth–twentieth century.

---

[6] Thomas L. Hankins, "In defence of biography: the use of biography in the history of science," *History of Science*, 1979, Vol. 17, 1–16; for quote see 5. For a more recent discussion of the role of biography in the history of science see L. Pearce Williams, "The Life of Science and Scientific Lives," *Physis*, 1991, No. 28, 199–213.

[7] A large number of recent works have revitalized and reinvigorated both the intellectual and historiographic significance of scientific biography, a few of which are: Nathaniel Comfort, *The Tangled Field: Barbara McClintock's Search for the Patterns of Genetic Control* (Cambridge, MA: Harvard University Press, 2001); Frederick L. Holmes, *Hans Krebs*, Vol. 1 (New York: Oxford University Press, 1991), xv–xx; Thomas Soderqvist, "The Seven Sisters: Subgenres of Bioi of Contemporary Life Scientists," *Journal of the History of Biology*, 2011, Vol. 44, No. 4, 633–650.

As a 'man of the eighties,' Vinogradskii represents a generation that bridged two periods of vigorous expansion in Russian science: the second half of the nineteenth century (the generation of the 'men of the sixties') and the early Soviet period. Vinogradskii lived through six decades of dramatic change in Russian and Western European history, and his career choices reflected the social, political, and intellectual forces that shaped these cultures.

The reader may at times be surprised by Vinogradskii's choices in mentor, method or conclusions, or commiserate with his frustrations. The struggle he endured represented a grand shift occurring in the history of the life sciences, in the changing relationship between natural history and laboratory research. His career exemplifies how scientists strove to balance their commitments, on the one hand, to romantic ideals associated with natural history, and on the other, with the new techniques and escalating status of laboratory-based investigations. How to preserve the beauty, and as we will see in Vinogradskii's reports, poetry of nature, in the new language of experiment, instruments, rigor, genuine or rhetorical? In the first half of the nineteenth century, efforts to categorize nature had slowly given way to increasingly dynamic natural historical systems including Humboldt's phytogeography, Lyell's historical geology, and Darwin's theory of evolution. The foundation of Vinogradskii's story occurs, however, in the second half of the century, which witnessed an increased reliance on laboratory research. With its focus on the chemical and physical investigation of organic and inorganic bodies, and on the experimental ideal of knowledge, this challenged the validity of the natural historical approach.

The laboratory revolution was neither a paradigm shift nor a changing of the guard. It was, rather, a period of slow transition marked by the blending of traditions. Vinogradskii's negotiation of these changes was reflected in his choices among disciplines, methodologies, and scientific questions. His excitement in the new approach, tempered by a reluctance to relinquish the past, sounded a quavering note as he first stepped onto the long road toward synthesizing these approaches. His commitment to the cycle of life, theoretical vision rooted in natural history, was strong yet flexible enough to absorb his penchant for experiment. That is, he learned how to express the concept of the cycle of life in the language of the laboratory—in the language of microscopic observations, chemical analyses, and gel plates. And on the some tone, he introduced the analytic, dissecting, and observing power of the laboratory into the romantic wild of nature.

In recent years, historians of science, who had largely neglected soils science (leaving it to practitioners to tell) have improved their efforts and achievements in this regard—Vinogradskii's story is my contribution. Through his biography, I begin to explore the growth of this science at the turn of the century when it was first becoming an established discipline. Introducing the concept of autotrophism and the elective culture method into soil science, Vinogradskii brought that science into contact with microbiology. As these disciplines expanded and divided into subfields in the twentieth century, his contributions facilitated their transformation into ecological sciences.

From a distant perspective, Vinogradskii's transformation was from a botanist into an ecologist—but the steps within are crucial. In Part I, I explore Vinogradskii's apprenticeship at St. Petersburg University, which introduced him to the scientific theories and methodologies that would shape his research over the next 60 years. Here he conducted his first investigation, applying a synthesis of techniques from Russian plant physiology and Pasteurian microbiology to questions of fungal nutrition. Vinogradskii's commitment to a thermodynamic view of nature known as the cycle of life—which depicted nature as a holistic process of transformation of matter and exchange of energy—underlay his investigation.

In Part II, we find that during an apprenticeship with Anton De Bary in Strassburg, Germany, Vinogradskii used the experimental techniques he had learned at St. Petersburg University to study new questions of microscopic fungal nutrition. This work reflected Vinogradskii's lasting commitment to both the cycle of life concept and a novel evolutionary perspective that was emerging during the 1880s in German botany. Although he only rarely discussed evolutionary theory, Vinogradskii's rejection of Darwinian evolution places him within the cadre of Russian biologists who nevertheless were responding to, and thus were influenced by, the new evolutionary perspectives and ongoing research. Ecology did emerge in part from evolutionary theory, but not in the limited ways discussed in much of the literature of history of ecology.

Part III describes Vinogradskii's move to Zurich in 1888, where he broadened the scope of his research program to include new organisms. Based on investigations of nitrogen bacteria, which reinforced his previous conclusions about the nature of microorganism nutrition, he introduced a new physiological definition for vital activity. His conclusion that a selection of living organisms could subsist solely on inorganic matter reflected his vision of the cycle of life and his ability to investigate that concept experimentally. His discovery of autotrophism attracted the attention of the scientific community and several job offers. Accepting the offer to join the recently founded Imperial Institute of Experimental Medicine in St. Petersburg, he once again entered a new institutional context. As director of the Institute's division of General Microbiology he continued his research on nitrogen bacteria research, addressing fundamental questions related to bacterial nutrition and applied medicine. For 15 years until his retirement in 1912, he applied new 'elective culture' techniques in his research, outlining specific nutrient cycles within nature's grand cycle of life.

In Part IV, we find Vinogradskii fleeing Russia after the Bolshevik seizure of power, and resettling in Brie-Comte-Robert, France, where he rekindled his investigation of nature's cycle of life. In this new French context—marked by its preference for ecological and applied science—he adapted his earlier research program to the vogue for agricultural science and soil microbiology. Transforming the cycle of life into an ecological perspective, he developed and promoted a new experimental method that gained the support of a wide international community of soil scientists. Through this international network Vinogradskii's new conceptualization of the cycle of life entered a broad range of fields including the biogeography of Vladimir Vernadsky, and the ecological microbiology of Rene Dubos and Selman Waksman.

Vinogradskii's efforts at Brie-Comte-Robert to push soil microbiology in a new direction found an accepting international audience. This warm reception was based in part on the abiding significance of his earlier discoveries in the late nineteenth and early twentieth century, especially his discovery of autotrophism (or chemosynthesis) and investigation of global nutrient cycles. By the 1930s, although this work still captured the imagination of plant physiologists and microbiologists, his reincarnation as a "French ecologist" attracted an expanded network of scientists.

In Part V, I survey the uptake of Vinogradskii's methods. His experimentalist-ecological approach in microbiology and soil science appeared just as ecology was becoming established as a discipline. Many soil scientists, soil microbiologists, and forestry and marine scientists found much of value in his laboratory methods. I discuss five institutional settings from quite varied regions of the world, where his methods had significant influence: at the Delft School of Microbiology in Amsterdam; the Agricultural Experiment Stations at Rutgers University and Roth Amsted, England; and at the Department of Agricultural Microbiology, Leningrad. Although Vinogradskii's work attracted the greatest interest among those who studied the soil ecologically, we will see that they assimilated his methods according to the demands of their own experimental programs.

In sum, through Vinogradskii's story I tell a history of microbiology that takes account of soil science and ecology. Yet what follows is also a history of soil science from the perspective of ecology and microbiology. Thirdly, I present a history of ecology as found in microbiology and soil science. Sergei Vinogradskii's biography unites these histories in a unique way—through the concept of the cycle of life.

Part I
Plant Physiology

Chapter 1
A Synthesis of Thermodynamics and Bioenergetics in Plant Physiology: The Investigation of a Moody Apprentice

On a dark and wintry day in 1883, the St. Petersburg Society of Naturalists met to hear the conclusions of a 3-year investigation in plant physiology. To grow normally, the audience learned, fungi require magnesium and an abundance of oxygen. This news, likely seen by the Russian naturalists in attendance as a mundane contribution to plant physiology, meant much more for its bearer, Sergei Nikolaevich Vinogradskii. His oral presentation marked the completion of a long and tedious apprenticeship, one he had entered into reluctantly at best, and which offered him no apparent job opportunities. Depressed and frustrated by his seemingly futile efforts, he returned home to his Ukrainian estate in an attempt to leave as much of this experience behind him. The investigation, however, had left an indelible mark on Vinogradskii's intellectual profile.

Unbeknownst to him, the questions he asked in this early experiment, and the methods by which he explored them, would inaugurate his 60-year career of microbiological research, eventually attracting an international circle of disciples. His investigation of 1880–1883 provided evidence to support an innocuous claim—only certain, identifiable environments would cause the microscopic organism, *Mycoderma vini Desm.* to develop its characteristic structure and patterns of growth. *Mycoderma* was not an exotic organism—unwittingly brewers of beer, wine, and vinegar had domesticated it for millennia in their fermentation processes. Once *Mycoderma's* role in fermentation had been discovered, however, they came to occupy a central position in numerous controversies concerning spontaneous generation, and the biological and chemical definitions of fermentation.[1] As late as the

[1] In 1826, Desmazières, a botanist in Lille divided the genus *Mycoderma* (coined by C. J. Persoon in 1822) into several species, including *M. vini*. In 1836–1837, three scientists (Baron Charles Cagniard-Latour, Theodor Schwann, and Friederich Traugott Kutzing) independently discovered the biological "fermentative" nature of these "yeast" cells. Three chemists (J. J. Berzelius, F. Wohler, and Justus Liebig) challenged this biological definition of fermentation, offering their own chemical definition. As influential as Pasteur's works on fermentation in the 1860s and 1870s

L. Ackert, *Sergei Vinogradskii and the Cycle of Life: From the Thermodynamics of Life to Ecological Microbiology, 1850-1950*, Archimedes 34,
DOI 10.1007/978-94-007-5198-9_1,

1880s, researchers were divided even over naming the globular *Mycoderma* cells, using the terms: brewer's yeast, *Zuckerpilz* (sugar fungi), bacteria, and *Spaltpilze* (fissionable fungi). The general disagreement about this organism signified its place at the frontier of science. Like all microorganisms, it existed at the limits of contemporary microscope technologies, chemical analytical methods, and scientific and popular imagination.

These exciting implications enveloped Vinogradskii's straightforward description of his morphological and physiological observations. In order for *Mycoderma* to grow into a luxurious film, for example, it required a culture containing magnesium and either zinc salts, potassium, rubidium, potassium chloride, or sodium chloride. Other compounds, like cesium or calcium, however, had no effect on *Mycoderma* growth. An abundance of oxygen, his investigation showed, caused *Mycoderma* to exhibit typical budding; in an oxygen-poor environment, the fungi formed mycelia. These esoteric conclusions were the results of nearly 3 years of labor devoted to creatively designing laboratory apparatuses, to producing elusive sterile cultures of microscopic organisms, and to observing their development in series upon series of nutritive solutions. His results survive only in relatively obscure form in a three-page summary of his presentation to the St. Petersburg Society of Naturalists. Written in terse Russian, this review, however, does point to Vinogradskii's involvement in several critical issues in contemporary physiology and microbiology.

Vinogradskii's exploration of fungal nutrition placed his investigation within plant physiology, an emergent field that strove to synthesize the taxonomic and morphological aspects of botany with a laboratory-based physiological approach. By the last quarter of the nineteenth century, the practices and ideals of physiology—featuring the "quantitative delineation of organic phenomena, experimental control over those phenomena, and aspiration toward prediction of phenomena"—were being extended to many domains of biology, including botanical research.[2] These developments were informed substantially by diverse conceptualizations of the conservation of forces and energy and the rise of thermodynamics in the mid-nineteenth century.[3] Thermodynamics at this time was an integral part of physiology, especially in the investigations of heat and the transformation of energy

were, they failed to convince many of the biological nature of fermentation, including Liebig, Oscar Brefeld, and Moritz Traube. See William Bulloch, *The History of Bacteriology* (London: Oxford University Press, 1938), 47–63; and Joseph S. Fruton, *Molecules and Life: Historical Essays on the Interplay of Chemistry and Biology* (New York: Wiley-Interscience, a Division of John Wiley and Sons, Inc., 1972), 42–66.

[2] William Coleman, *Biology in the Nineteenth-Century: Problems of Form, Function, and Transformation* (New York: Wiley, 1971), 162.

[3] There is an extensive literature on the history of the conservation of energy and thermodynamics. Fabio Bevilacqua, "Helmholtz's *Ueber die Erhaltung der Kraft*: The Emergence of a Theoretical Physicist," in David Cahan, ed., *Hermann von Helmholtz and the Foundations of Nineteenth-Century Science* (Berkeley: University of California Press, 1990), 291–333; Kenneth L. Caneva,

(metabolism) in animals. By the 1870s, however, experimental physiologists recognized that they had reached a methodological impasse. In the 1850s, Justus Liebig, Hermann Helmholtz, and Robert Mayer conceptually had transformed the static chemical methods of Jean-Baptiste Dumas and Jean-Baptiste Boussingault into a thermo-"dynamic" method. The principle of the conservation of energy as conceived by Liebig, Helmholtz, and Mayer, however, "black boxed" the physiological processes occurring in living organisms, measuring only input (food) and output (changes in heat).[4] By the 1880s, the application of the conservation of energy to physiology had raised a new set of theoretical and experimental issues that required a new methodology. To investigate the vital processes occurring within organisms at the molecular and cellular level—inside the "black box" of life—physiologists drew on the new chemical approach known as "bioenergetics." For them this was a departure from the thermodynamics of life, which studied the energy exchanges at the level of the whole organism.[5]

The new divide between physiological bioenergetics and the thermodynamics of life did not exist for microbiologists, whose single-celled organisms challenged the basic demarcation between the cellular and the whole organism. These microbiologists adopted the bioenergetics approach yet retained the thermodynamic conceptualization of nature, developing holistic visions in which nature's inorganic and organic realms were united by cyclical exchanges of matter and transformations of energy.

Robert Mayer and the Conservation of Energy (Princeton: Princeton University Press, 1993), 160–206; Yehuda Elkana, *The Discovery of the Conservation of Energy* (London: Hutchinson Educational, Ltd., 1974), 114–145; Peter M. Harman, *Energy, Force, and Matter: The Conceptual Development of Nineteenth-Century Physics* (Cambridge: Cambridge University Press, 1982), 60–64; Aaron J. Ihde, *The Development of Modern Chemistry* (New York: Harper & Row, 1964), 395–399; A. J. Kox and Daniel M. Siegel, eds., *No Truth Except in the Details: Essays in Honor of Martin J. Klein* (Dordrecht: Kluwer Academic Publishers, 1995, see especially section three "Thermodynamics and Matter Theory, Physical and Biological," 135–244; Thomas S. Kuhn, "Energy Conservation as an Example of Simultaneous Discovery," in Marshall Clagett, ed., *Critical Problems in the History of Science* (Madison: University of Wisconsin Press, 1969), 321–356; Ernst S. C. von Meyer, *A History of Chemistry: From the Earliest Times to the Present Day*, trans. George McGowan (New York: Macmillan and Co., Ltd., 1898) Second English Edition, 507–509, 530–551; Robert D. Purrington, *Physics in the Nineteenth Century* (New Brunswick: Rutgers University Press, 1997), 105–112; Anson Rabinbach, *The Human Motor: Energy, Fatigue, and the Origins of Modernity* (Berkeley: University of California Press, 1990), esp. chapter two "Transcendental Materialism: The Primacy of Arbeitskraft (Labor Power)," 45–68; Alan J. Rocke, *Chemical Atomism in the Nineteenth Century: From Dalton to Cannizzaro* (Columbus: Ohio State University Press, 1984), 288–311; David W. Theobald, *The Concept of Energy* (London: E. and F. N. Spon, Ltd., 1966) 145–159.

[4] On the impact of thermodynamics on biology and physiology see Richard L. Kremer, *The Thermodynamics of Life and Experimental Physiology, 1770–1880* (New York: Garland Publishing, Inc., 1990), 453–455; and Timothy Lenoir, *The Strategy of Life: Teleology and Mechanics in Nineteenth Century German Biology* (Dordrecht/ Holland/Boston: D. Reidel Publishing Co., 1982), 197–215, 229.

[5] Kremer, *The Thermodynamics of Life and Experimental Physiology, 1770–1880*, 16, 23–25.

Pasteur's *cycle de vie*, Ferdinand Cohn's *Kreislauf des Lebens*, and Andrei Famintsyn's *obmen veshchestva* (only a few of similar 'cycle of life' concepts) at once preserved the conservation of energy in their research, as well as distantly reflecting Romantic *Naturphilosophie* visions of the interconnectedness of the cosmos. After the 1870s, however, these scientists strove to express their visions using the new chemical and physical experimental methods of their day. Debates about the 'thermodynamics of life' raged on during Vinogradskii's apprenticeship and provide part of the context for his methodological choices and interpretations.[6]

Vinogradskii's investigation placed him at the very center of these methodological and theoretical crosscurrents in the international science of his day. The ideas and methods of these Western European schools, although geographically peripheral to his own Russian setting, were available to him through the teachings of his mentors—who had studied in Western Europe—and in the locally-available, foreign scientific literature. He benefited from this international educational exchange during his studies with the botanist Andrei Famintsyn at the University of St. Petersburg. Under Famintsyn's mentorship, Vinogradskii applied his new experimental skills in a research project aimed at European-wide controversies in plant nutrition and the nature of microbial life.[7] His 1883 summary of this research reveals not only his position in these debates, but also his methodological preferences for investigating that nature. Understanding these views and approaches also provides a foundation for following later developments in his research.

The personal side of this story is inseparable from the scientific. Like the biological sciences of the 1880s, Vinogradskii was experiencing his own growing pains. Born into the landed gentry, he was independently wealthy. Yet, according to the mores of his class he was driven to contribute to the newly emerging society.[8]

---

[6] I borrow the term 'thermodynamics of life' from Kremer, who did not include Boussingault and Dumas in his study because they had considered neither heat nor work, per se, in their research. They were, however, major players in the history of the cycle of life concept, providing the foundations for Pasteur's, Felix Hoppe-Seyler's, and Famintsyn's research; see Ibid., esp. Chapter Six "Animal Heat and Energy Conservation, 1837–1847," 190–259. For a discussion of the rise of thermodynamics and the concept of energy as it applied to these changes in physiology in the mid-nineteenth century see: Eduard Glas, *Chemistry and Physiology in Their Historical and Philosophical Relations* (Delft: Delft University Press, 1979); Frederick Lawrence Holmes, *Claude Bernard and Animal Chemistry: The Emergence of a Scientist* (Cambridge, MA: Harvard University Press, 1974), see esp. the conclusion 445–455; George Rosen, "The Conservation of Energy and the Study of Metabolism" in Chandler McC. Brooks and Paul F. Cranefield eds., *The Historical Development of Physiological Thought* (New York: The Hafner Publishing Company, 1959), 243–263; and; Mikulas Teich, "The Foundations of Modern Biochemistry" in Joseph Needham, ed., *The Chemistry of Life: Eight Lectures on the History of Biochemistry* (Cambridge: Cambridge University Press, 1970), 171–191.

[7] The Russian instructors should not be seen as only receivers and interpreters of this knowledge; many of them (Famintsyn, Borodin, and Beketov, for example) had studied in the European schools and laboratories and had made lasting contributions. They also continued to publish their Russian research in the European-based journals.

[8] See Kendall E. Bailes, *Science and Russian Culture in an Age of Revolutions: V. I. Vernadsky and His Scientific School, 1863–1945* (Bloomington: Indiana University Press, 1990), 15–17.

In search of a "career path" he was struggling to choose between three possible lives: as *pomeshchik* (landed-gentry 'farmer'), a virtuoso pianist, or a 'professor' in the natural sciences.[9] The young man was torn between three loves—family, music and this new fascination with science. Eventually deciding for the professor's track, he experienced an increasing demand on his time and energy. The new pressures of setting up and running sensitive and protracted experiments clashed with his ongoing and unavoidable responsibilities to his growing family—he had recently married and they had their first of four daughters—to the sprawling family estate near Kiev, and with his passion for the forte piano and violin. These competing pressures tore at Vinogradskii's psyche during the course of his apprenticeship, and upon its completion he found himself living at the family estate, playing informal concerts with his relatives, and tinkering alone in his homemade laboratory—a virtuoso of nothing.

The Gentleman Chooses Science

Vinogradskii's status of amateur naturalist after his demanding apprenticeship stemmed not only from personal and social pressures, they also stemmed from his gymnasium years. His initial interest in and commitment to the natural sciences had occurred during a period of great personal indecision.[10] The substantial wealth and property his father had amassed as president of the first Bank of Commerce in Kiev and from his investments in beet sugar production placed the Vinogradskii family in the upper crust of the landed gentry. These family riches, however, did little to satisfy the 17-year-old Sergei Nikolaevich, who, looking down the long empty road of his future upon graduation, felt in dire need of advice and counseling. He felt abandoned at this critical juncture in his life.[11]

---

[9] Sergei N. Vinogradskii, *Itogi (In the End)*, these unpublished reminiscences were most likely written in the 1940s. They are located in the Arkhiv Rossiiskoi Akademii Nauk, Moskovskii Filial, fond 1601, opis' 1, delo 13. When Vinogradskii referred to "professor' here he was thinking along the lines of a Dmitri Mendeleev—a professor who taught courses, and conducted research in their own laboratory.

[10] Vinogradskii, *Itogi*, 1.

[11] Biographical treatments of Vinogradskii include: Vasilii L. Omelianskii, "Sergei Nikolaevich Vinogradskii: Po povodu 70-letiia so dnia rozhdeniia," *Archiv Biologicheskikh Nauk* 28 (1927), 11–33; and Selman A. Waksman, *Sergei Winogradsky: His Life and Work* (New Brunswick: Rutgers University Press, 1953). Recently a new work appeared containing the above works by Omelianskii and Waksman; see Iu. A. Mazing, T. V. Andriushkevich, and I. P. Golikov, eds., *Rasskazi o velikom bakteriologe S. N Vinogradskom* (St. Petersburg: Rostok, 2002). I also draw on his reminiscences *Itogi*. I acknowledge that Vinogradskii's portrayal of himself (his 'life's film' as he called it) should not be accepted entirely at face value; his self-representation must be interpreted in light of its own particular context. Vinogradskii was nearing the end of his life, he felt abandoned by his daughter Helen Vinogradskii, and Waksman was persistently encouraging him to send autobiographical materials. Vinogradskii, who was more than a little embarrassed about having a biography written about him while he was still living, resisted Waksman's requests. (See Vinogradskii, *Itogi*, 3–4).

Vinogradskii remembered this period as being difficult and trying. Assessing possible career options, that of a government official, a virtuoso pianist, or an accountant, he wrote: "the film of my life completely and sharply deviated from those predominant types [of lives] in nearly all points."[12] He remembered himself being "a humble, disciplined boy" who "passed his gymnasium years with success" and who "from the beginning to the end was a star student who had earned a gold medal."[13] At the time, however, he took no pleasure in his successes—he "hated his gymnasium and did not value the reward from it."[14] Upon graduation from this "loathsome" school, he "stood face to face with the question of what to do." He might have followed his brother Alexander who was a year older and following the example of their father earlier, had entered the juridical department of the University of Kiev to prepare for civil service.

The quandary situation experienced by the Vinogradskii brothers, exemplified the predicament of their generation and social class. In the 1860s, a new social structure was emerging in post-emancipation Russia. The gentry, the class to which the Vinogradskiis belonged was in decline and leading adolescents to seek out new professional niches.[15] Contemporary writers and later social scientists most often portrayed this generational divide as a rebellion of sons who came of age during the 1860s and 1870s and disregarded their fathers' values. One facet of this rebellion, and the broader ongoing social transformation, was a dramatic increase in interest in science, which was also garnering substantial governmental support.[16] In Vinogradskii's later opinion, his own cohort failed to join in this rebellion. Although they were dealing with the same questions, he approached life differently that his friends did—where they suffered, he thought, from a lack of "any strivings or interests," "he asked questions, hesitated, and searched."[17] The contrast he perceived between himself and his friends might be attributed to the personal dilemma that he was facing: the gymnasium had failed to inspire him in anything—it provided him "no interests, no strivings, and did not awaken any vocations." In addition, there was no advice to be found—his father was already seriously ill and his mother had

[12] Vinogradskii, *Itogi*, 1.

[13] Ibid., 2.

[14] Ibid., 3.

[15] These changes were the subject of Turgenev's, *Father's and Sons*, which is often taken as a reflection of the divide in generations over the place of science in Russian society. Turgenev's controversial analysis of Russian society has been attributed the formation of V. I. Vernadsky's political and social values. See Bailes, *Science and Russian Culture in an Age of Revolutions: V. I. Vernadsky and His Scientific School, 1863–1945*, 1–5; and Paul Miliukov, *Russian and Its Crisis* (London: Collier Books, 1962), 270–273.

[16] Daniel Todes treats these issues in his discussion of the Russian reception of Darwin's evolution by natural selection in Daniel P. Todes, *Darwin Without Malthus: The Struggle for Existence in Russian Evolutionary Thought* (New York: Oxford University Press, 1989), 20–23; and Alexander Vucinich, *Science in Russian Culture, 1861–1917* (Stanford: Stanford University Press, 1970), Vol. 2, 3–34.

[17] Vinogradskii, *Itogi*, 3.

neither any authority over him nor (because she was too busy with her new responsibilities) the inspiration to claim that authority. Even if he had found "someone [else] in the household or a friend of the family" who might "have taken to heart the career questions of a 17-year-old youth," he would probably have heard the same sterile advice: "follow the beaten path" and study a practical subject such as law or business.[18] Lacking a father or anyone from the older generation to react against, this son prolonged and stumbled over career decisions his peers made, if not thoughtfully, at least more quickly.

Vinogradskii did temporarily follow "the beaten path," and joined his brother at the juridical department. Unlike Alexander, who "immediately felt a lively interest in encyclopedic or governmental law," Sergei experienced "a deadly boredom" from his very first law lecture. His dread of this subject was so great it eventually fortified him enough to overcome his shyness, and fear of appearing "flippant" [*legkomyslennyi*], and he petitioned to transfer into the natural sciences department.[19] Why he chose the natural sciences are unclear. Perhaps his decision was, per Sulloway's model, a manifestation of rebellion often associated with later-born siblings.[20] The natural sciences—although at that time it was becoming a respected activity in Russian intellectual society—did still enjoyed little prestige in Vinogradskii's provincial society.[21] His resolve to study the "real sciences" was strong enough to overcome not only this perceived strangeness by his peers and elders, but also "his own complete lack of information about even their very rudiments."[22]

The experience disappointed him: the limited science curriculum lacked vigor and attracted very few students.[23] Thus, ignoring his family's protests, in his third year Vinogradskii applied to the St. Petersburg Conservatory.[24] There he would study with Theodor Leschetizky whose novel, "modern" methods of piano instruction

---

[18] Ibid.

[19] Ibid., 5.

[20] For a discussion of this topic, see Frank J. Sulloway, *Born to Rebel: Birth Order, Family Dynamics, Creative Lives* (New York: Vintage Books, 1997).

[21] Vinogradskii, *Itogi*, 4 oborot.

[22] Ibid.

[23] Waksman bases these comments on oral interviews of Vinogradskii, on their correspondence and possibly also in a now lost informal, 700 page autobiography by Vinogradskii. See Waksman, 8–9.

[24] Later Vinogradskii felt that music had prohibited him at this early stage to develop an attraction to the natural sciences and "in the end even paralyzed it." See *Itogi*, 15. That the family would protest is somewhat surprising, since an interest in music prevailed in the Vinogradskii home; the matriarch, Natalia V. Skoropadskaia, had taught her sons music at home. Perhaps the family was just concerned that he kept switching career paths. See *Itogi*, 4. There was also an O. Vinogradskii involved in the Russian Imperial Music Society of Kiev. Selman Waksman relates that Vinogradskii's brother Alexander, while studying law at the University of Kiev, also "became intensely interested in music." See Waksman, *Sergei Winogradsky: His Life and Work*, 7.

were attracting students from around the world.[25] Relative to acceptance by the Conservatory, for Vinogradskii "everything else paled, and lost its fragrance;" his new goal in life became "making himself into a musical artist—a virtuoso."[26] For the first time in his life he felt in control and independent; "no one held him back ... no one subjected him to criticism."[27] This new, exciting life, he would soon learn, had its tedious side. He attended lectures and took his exams, but without any "vital spirit." The initial pleasure of his decision, like everything else, soon waned. Self-criticism shadowed him during his "very ordinary beginning" in musical work. Increasingly, though, Leschetizky's dynamic personality and the novelty of his methods reinvigorated Vinogradskii's interest in the piano.

Ultimately, not even the masterful Leschetizky could mold a genius from Vinogradskii's mundane clay. During annual evaluations the Conservatory deemed that although Vinogradskii had talent, "it was the talent of a Saliari, and a long way from that of a Mozart."[28] Vinogradskii had worked zealously, he later recalled, but "without that fire of the unconscious or subconscious inspiration that characterizes genuine artistic natures."[29] His decision to leave the Conservatory, whether forced by its "authorities" or initiated himself, was for Vinogradskii the only possible recourse. Sixty years later, reminiscing in his French "castle" (likely casting a glance toward his well-worn piano that he still played daily), the sting of his failure had faded enough that he could write that he had abandoned the conservatory "without regrets."[30] The bitter disappointment, no doubt, was fresh in his mind that fall when he matriculated at St. Petersburg University.

Vinogradskii described his return to university studies as "imperative" and "penitential."[31] Although he had completed his junior year at the University of Kiev, St. Petersburg University placed him only in the sophomore year. Distrusting his own general abilities and level of knowledge, he attended lectures zealously, hoping to

---

[25] Leschetizky taught at the Conservatory between 1867 and 1879. Vinogradskii studied with him for 1 year and 3 days (16 September 1876–19 September 1877) not enough time to be considered his student, but long enough for music to become a steady companion in his life. Some of Leschetizky's other students recalled that he considered a person his "student" only when that person had studied with him for at least 2 years. See the biographies on Theodor Leschetizky at the St. Petersburg Conservatory see Tsentral'nyi Gosudarstvennyi Istoricheskii Archive, St. Petersburg, fond 361, opis' 9, delo 10, No. 90, 3 lista— "Lichnoe delo Professora T. Leshetitsogo." On Sergei N. Vinogradskii at the Conservatory, see idem. fond 361, opis' 1, delo 693, 5 listov—"Vinogradskii, Sergei, 16/9/1876 – 9/9/1877." On Alexandr Nikolaevich Vinogradskii's time at the Conservatory see fond 361, opis' 1, delo 362, 1 list— "Vinogradskii, Alexandr 5/1/1879 – 20/1/1879." Vinogradskii later recalled that one of the reasons he quit music and returned to study the natural sciences was that Leschetizky had left Russia. These dates, however, seem to tell another story.

[26] Vinogradskii, *Itogi*, 5.

[27] Ibid.

[28] Ibid.

[29] Ibid.

[30] Byron H. Waksman (Selman Waksman's son), who had visited Vinogradskii at Brie-Compte-Robert, France in the mid 1940s witnessed Vinogradskii's playing. Byron's impression was that Vinogradskii's adept skill at the piano was recognizable even through the clumsiness of his nearly 90 year old fingers. Oral interview with Byron H. Waksman (April 2001).

[31] Vinogradskii, *Itogi*, 6.

make up the lost ground. He originally intended to study chemistry, and enrolled in Dmitri Mendeleev's course on inorganic chemistry. Mendeleev was enjoying great fame as the creator of the periodic system of elements and his lecture hall "was always full: probably with students from other departments who came to the lectures to gaze at Dmitri Ivanovich, at his large slovenly figure and his long hair, beard to the floor, and to hear how a celebrity speaks." Whether it was the poor conditions of the lectures:

> they met "every day early in the morning at 9 am, and maybe even at 8 am, when it was still dark in Petersburg; the darkness was increased by [thick] forest of trees that screened the main facade, onto which the large, gloomy auditorium looked out

or Mendeleev's delivery:

> he spoke like people speak who are not sufficiently garrulous, who do not take any pains to finish their own lecture

for Vinogradskii "the hours dragged along endlessly."[32] Catching himself "dozing off heavily and invincibly," he wondered: "am I powerless to pay attention and grasp the chemistry being taught by an acknowledged great scientist?"[33] Mendeleev did deliver at least two "striking" lectures, Vinogradskii recalled: "when he explained the gradations of scientific development to us: hypotheses grow into theories and consolidate in science as laws," and "on the periodic system of elements, which we as beginners could not understand very well."[34] In general, however, Vinogradskii considered Mendeleev's "boring" lectures a great waste of time.[35]

Nor did he find other courses in chemistry more engaging. The mumbling, gray-bearded genius Mendeleev is perhaps too easily blamed for Vinogradskii's apathy towards chemistry. Although he did better in other chemistry courses during less gloomy times, overall he still found the subject completely unfamiliar, alien, and fatiguing.[36] In sharp contrast, Andrei Famintsyn (Mendeleev's unwitting competitor for Vinogradskii's attention) prepared his botany and plant physiology lectures "carefully and conscientiously" and delivered them "smoothly, clearly and not

---

[32] He further described Mendeleev's style: "In general (crossed out—the speech went somewhat) he spoke slowly, stopping himself and searching for expressions, while often dwelling and dragging for along time and speaking in a deep voice [басисть] um-um-um-um, says a work and then a second time um-um-um. Excessively long. These mumblings and in general this character of exposition did not electrify the audience very much." Vinogradskii, *Itogi*, 7. Mendeleev was so infamous for his torturous lecture style that Michael Gordin, *A Well-ordered Thing: Dmitrii Mendeleev And The Shadow Of The Periodic Table* (Basic Books, 2004) at Princeton University is collecting anecdotes describing it.

[33] Ibid.

[34] Ibid., 8.

[35] Ibid.

[36] Later, after his experiences in the German school system, and through his daughter Elena's experience at Cambridge in England, Vinogradskii would blame "not the professors, not even the mumbling of Mendeleev—and not the bluntness of the students, but rather the lecture system itself." "In Germany, he wrote, lecture courses were also delivered—(or to be more exact, were delivered in my time)—in all sciences, but the student was able to choose the courses that most interested him. [There were also] the secondary subjects for him, for example: botany and chemistry, zoology and chemistry, chemistry and physics etc.; and thus one had to attend not more than one or two lectures per day." Ibid., 7–8.

monotonously."[37] Aside from Famintsyn's youthful charisma and lecture style, he offered an exciting combination of observational and experimental approaches to the study of nature that renewed Vinogradskii's interest in science.[38]

The botanical department is centrally isolated on St. Petersburg University's campus in "a two story building in the heart of the University courtyard." Distant from the other buildings it offered "a perfectly separate little world."[39] Andrei Beketov, the department chair, managed his herbarium on the first floor, and on the second Famintsyn taught plant anatomy and plant physiology.[40] Beketov and Famintsyn shared a common, effective, approach to instruction. They offered initial discursive lectures that served as preface or commentary to practical, hands-on lessons that followed.[41] Though Vinogradskii did obtain extensive training in natural history from Beketov, his approach was less exciting than Famintsyn's. Beketov described "the characteristics of several families and their most interesting representatives … while [his students] sat at a table in front of the windows with magnifying glasses independently studying the herbarium exemplars to define the species of those plants."[42] Vinogradskii soon graduated from Beketov's herbarium—with its magnifying glasses and dried specimens—to Famintsyn's physiology laboratory, with its "more interesting" microscopes, aquariums, and intricate glassware.[43]

Recognizing Vinogradskii's devotion to plant physiology (he regularly frequented the lab and had begun to acquire his own library of essential literature), Famintsyn soon accepted him as a *stagiaire* or trainee.[44] Whether or not his new authorities had distinguished him as a 'virtuoso in the rough,' Vinogradskii felt comfortable in the botanical department. He enjoyed the company and the support of both Famintsyn and his assistant P. Ia. Krutitskii. In his senior year, when students needed to elect their main and secondary scientific disciplines, he chose botany and zoology. Released from "sitting through long hours of lectures [he now] spent all [his] time

---

[37] Ibid., 9.

[38] Later Vinogradskii would add that "since the experimental sciences are completely based on observations and experiments, the main method to introduce beginners in any of the sciences is … only by way of observations and experiments, which should be carried out in parallel with theory, and best of all if a little bit beforehand; this is in order that the participant can familiarize himself, if only a little, with the materials before hearing in detail the theory of a given field." Ibid.

[39] Ibid.

[40] On Beketov see B. P. Strogonov, *Andrei Sergeevich Famintsyn, 1835–1918* (Moskva: Nauka, 1996), 25–26; Todes, "Beketov, Botany, and the Harmony of Nature," *Darwin Without Malthus: The Struggle for Existence in Russian Evolutionary Thought*, 45–61.

[41] Vinogradskii, *Itogi*, 10.

[42] Ibid. Vinogradskii wrote: "Curiously, I wonder if these modest flora exercises attracted me to botany and in general served as a stimulus for independent studies of science. I continued them even during my vacations and attracted Momma, who did not give up, but surpassed me in familiarity of the local Podol' flora. I do not doubt that study of flora was the initial push of my scientific development."

[43] Ibid.

[44] Ibid.

in the laboratory in search of an interesting theme for independent work."[45] By the time he graduated from the University in 1880, Vinogradskii had found his path. He would train for a professorship in botany.

Next fall he took the first step along that road—an apprenticeship under Famintsyn that would lead to a Master's degree in botany. Vinogradskii imagined that the professorial track entailed: "joining the ranks of the university, completing a Masters examination, dissertation, then defense; possibly followed by securing a slot as senior lecturer (*dotsentura*); completing a Doctoral dissertation and defense; searching for a position; and finally developing courses."[46] Although he was unenthusiastic about this prospect, perhaps thinking back to his earlier moments of 'flippancy' he decided "to stick to the beaten path" and "began to assemble some independent erudition."[47]

A tension arose between Vinogradskii's newly discovered passion for independent research and his perception of Famintsyn's mentorship role. He fluctuated between enjoying his newfound independence and craving more guidance. On the one hand, he felt free to explore new techniques and lines of research by doing outside reading, and even successfully introduced these innovations into the economy of the laboratory. On the other hand, he exercised his independence primarily within the bounds of Famintsyn's mentorship: by taking Famintsyn's courses, pushing the limited resources of the laboratory, and responding to the inertia of the laboratory's investigatory direction and research interests. Vinogradskii's portrayal of Famintsyn's mentorship, informed especially by his earlier experiences at the Conservatory, reveals the dynamics underlying the tensions of his apprenticeship. 'Good,' for Vinogradskii, was not good enough; he needed to see himself as an innovator and a virtuoso. His respect for these qualities, demonstrated in his break from the Conservatory, in his behavior during the apprenticeship, and in his recollection of these episodes in 1949, influenced his investigatory choices not only while at Famintsyn's laboratory, but throughout his career.

Vinogradskii's description of his apprenticeship contradicts at times other evidence about Famintsyn's interests and skills. For example, he credited himself with introducing microbiology circa 1880, along with its newest problems and methodologies, to Famintsyn's laboratory. During his first year of apprenticeship, he recalled, he continued to expand his small library of scientific literature through regular visits to "Pikker's store on Nevsky where [he] looked over the news and in general subscribed to foreign literature."[48] He attributed to his independent reading the awakening of his interest in Pasteurian microbiology. "It is no wonder," he wrote, "that [Louis] Pasteur's works occupied the pivotal spot in my library," which "were at the time still novel."[49] Pasteur's *Études sur la vinaigre, Études sur la vin,*

---

[45] Ibid.

[46] Ibid.

[47] Ibid., 10–11.

[48] Ibid., 11.

[49] Ibid.

and especially his *Études sur la bière*, in which he outlined his theory of fermentation, became Vinogradskii's primary reference books.

These works of the late 1860s and early 1870s may have been relatively "novel" to some, but Famintsyn was quite familiar with them and had previously integrated Pasteur's ideas into his own research. Famintsyn had for example, completed numerous investigations testing an idea he had learned from studying Pasteur's crystallography work—"that at the foundation of life lies a force that resembles crystal-formation."[50] Famintsyn published his results in several articles on "Die Wirkung des Lichtes" (The Influence of Light) on various plant cells or single-celled organisms.[51] Famintsyn also esteemed Pasteur for his discovery of intramolecular respiration, a process during which plants separate carbonic acid (CO_2) without absorbing oxygen from outside their bodies.[52] Famintsyn not only drew on Pasteur's research in his own investigations, he also conducted a thorough survey of Pasteur's work for the St. Petersburg Society of Naturalists (1873) and taught Pasteur's work in plant physiology courses annually after 1875.[53] Famintsyn thus likely played albeit an unacknowledged role influencing his apprentice's great appreciation for the French scientist's work.

The technical conditions in the laboratory failed to meet Vinogradskii's "aspirations at any level" and he strove to remedy the situation.[54] He boasted of introducing to the laboratory, "the first retorts in the model of Pasteur," and that he built—although not very successfully—the laboratory's first thermostat.[55] We can take him for the most part at his word, however Famintsyn was also an accomplished craftsman. The latter had designed and built several complicated pieces of laboratory

[50] V. V. Polevoi, "A. S. Famintsyn i fiziologia rastenii v Peterburgskom-Leningradskom Universitete," A. L. Kursanov et al., eds., *Andrei Sergeevich Famintsyn: Zhizn' i nauchnaia deiatel'nost'* (Leningrad: Nauka, 1981), 56–85, see 66 for quote. On Pasteur's crystallography, see Gerald L. Geison, "From Crystals to Life," *The Private Science of Louis Pasteur* (Princeton: Princeton University Press, 1995), 90–110.

[51] For example see A. S. Famintsyn, Die Wirkung des lichtes auf Wachsen der Keimenden Kresse" *Mémoires de l'Académie sciences St. Petersburg*, Ser. 7, Tom 8, No. 15, 1–19; Idem, *Bulletin de l'Académie sciences St. Petersburg*, Tom 8, No. 3, 545–549; for Famintsyn's complete bibliography see Strogonov, *Andrei Sergeevich Famintsyn, 1835–1918*, 141–158.

[52] Eh. N. Mirzoian, "Evoliutsionno-biokhimicheskie vzgliady A. S. Famintsyn v sviazi s ego filosofskimi i obshchebiologicheskimi vozzreniiami," Kursanov et al., *Andrei Sergeevich Famintsyn: Zhizn' i nauchnaia deiatel'nost,'* 150–164, see 160.

[53] See B. P. Strogonov, *Andrei Sergeevich Famintsyn, 1835–1918*; for the listing of the 1873 synopsis of Pasteur's work entitled "Samobrozhenie plodov i pr.: O noveishikh pazyskaniiakh nad brozheniem" in *Trudy Sankt-Peterburgskogo Obschchestva Estestvoispytatelei,* see 144; on the plant physiology courses see 28–29.

[54] Vinogradskii, *Itogi*, 11.

[55] He ordered them "from Nitt's (on the Moika at the corner of Gorokovaia Street, at that time the only more-than-modest store of laboratory accessories)."Ibid.

equipment, which he had used effectively in his investigations of photosynthesis.[56] He was also quite used to working with microorganisms and had much of the necessary equipment, including microscopes, a variety of retorts and glassware, and the practical skills to use them effectively. During Vinogradskii's apprenticeship the technical conditions certainly were adequate enough for him to be the first to use a microscope camera to observe for a prolonged period the development of single isolated cells.[57]

Vinogradskii experienced a tumultuous emotional ride during his "student experiments."[58] He began to attract the attention of the university population, which fueled his "genuine passion" and:

> investigative ardor, when life moved from one experiment to the next and only in them did [one] find gratification. From them came all joys, and in them [lay] all hopes.[59]

Mixed with these joys and hopes, he regretfully noted, were "many moments of disappointment and depression."[60] The same drive for innovation that had encouraged him to explore the foreign scientific literature for new ideas and approaches also marked a divergence between his own interests and those of Famintsyn's laboratory. As Vinogradskii progressed in his microbiological research he felt increasingly isolated from the rest of the laboratory until, he felt, he was working in an atmosphere of "complete solitude."[61] Microbiology, in Vinogradskii's opinion, was "entirely unfamiliar" to the laboratory, including its chief, Famintsyn. Famintsyn's ignorance of microbiology, at least as Vinogradskii approached it, was not, however, Vinogradskii's sole reason for feeling isolated. He felt that during this period Famintsyn was too preoccupied writing his book *Obmen Veshchestvo i Prevrashchenie Ehnergii v Rasteniiakh* (*The Exchange of Matter and the Transformation of Energy in Plants*) to appear regularly at the laboratory.[62] The fluctuating joys and disappointments

---

[56] Famintsyn initiated this research in the mid 1860s at the same time as Julius Sachs had. Famintsyn's novel contribution was to use artificial light produced by kerosene lamps (in part due to the rarity of sunlight during the St. Petersburg Winters), which allowed him to more accurately measure the amount of energy being used by the plants studied. See A. S. Famintsyn, "Deistvie sveta kerosinovoi lampy na *Spirogyra orthospira Naeg*.," in *Deistvie Sveta na Vodorosli i Nekotorye Drugie Blizkie k nim Organismy* (St. Petersburg, 1865), 39–56. See also, E. M. Senchenkova, "Issledovaniia A. S. Famintsyna po Fotosyntezu" in Kursanov et al., *Andrei Sergeevich Famintsyn: Zhizn' i nauchnaia deiatel'nost'*, 86–109; on Sachs and Famintsyn's race see 88–92; and on the kerosene lamp work see 91–94, and Strogonov, 39–41.

[57] Strogonov, *Andrei Sergeevich Famintsyn, 1835–1918*, 63. There may have been, however, other, more obscure materials required by Vinogradskii's research, of which we have no evidence.

[58] Vinogradskii recalled being excited at this new attention attributing it to "the self-initiative of his undertaking." Vinogradskii, *Itogi*, 11.

[59] Ibid.

[60] Ibid.

[61] Ibid., 12.

[62] Ibid. A. S. Famintsyn, *Obmen Veshchestvo i Prevrashchenie Ehnergii v Rasteniiakh* (Sankt-Peterburg: Imperatorskoi Akademii Nauk, 1883).

Vinogradskii experienced during his experimental work so exacerbated his conflicting needs for both independence and guidance that he characterized this time as his "*Sturm und Drang* period."[63]

Vinogradskii's sense of complete isolation reflected the pressure he was feeling to innovate and succeed. The very book that Vinogradskii blamed for distracting Famintsyn from the laboratory offers evidence that his mentor was indeed quite knowledgeable about microbiology. Famintsyn's discussion of the debates surrounding alcohol fermentation, microbial taxonomy, and the proper methods for investigating Pasteur's "*infiniments petites*" indicate that his earlier interest in Pasteur's work continued to be one of his intellectual concerns.[64] It is significant that Vinogradskii's period of storm and stress coincided with Famintsyn's final push to complete *The Exchange of Matter* in 1881 (although it was not published until 1883).[65] If Famintsyn had been too busy preparing his monograph for the press, however, others in the laboratory were versed in his style of plant physiology, including the assistant Krutitskii.[66] Although it did not involve microorganisms, Krutitskii's research did require the techniques and skills needed to study plants at the cellular level. In addition, Famintsyn had organized a *kruzhok* (circle) of young botanists that Vinogradskii might have joined to share in discussions over tea or, on more adventurous days excursions. It is unclear whether his absence at these gatherings reflects antisocial behavior, or his increasing duties to his family. If he had genuinely craved intellectual stimulation and guidance, however, he could certainly have found it at these meetings and in the laboratory.

In the spring, as Vinogradskii left the cozy botany building and strolled along the block-long main hall of the university towards the great Neva River, and as his thoughts turned to spending the upcoming summer in rural Kiev, the life of a gentry farmer didn't seem so bad after all. He had suffered enough for science. Set adrift by his mentor, Vinogradskii had "meandered unsystematically solving no specific problem."[67] Looking back on this time, he described the complicated interconnections

[63] Ibid.

[64] Famintsyn, *Obmen Veshchestvo i Prevrashchenie Ehnergii v Rasteniiakh*, 1883. This book was republished under the same name as Idem (Moskva: Nauka, 1989), A. L. Kursanov, ed. My page numbers refer to the 1989 edition; it is a faithful reprinting and widely available.

[65] Strogonov, *Andrei Sergeevich Famintsyn, 1835–1918*, 131.

[66] P. Ia. Krutitskii was investigation the possibility of using the cellulose membranes in the stalks of *Phragmites communis* in diosmosis research. He would present his results at the same meeting of the Botanical section of the St. Petersburg Society of Naturalists, when Vinogradskii presented his own investigation, in December 1883.

[67] Ibid., Here Vinogradskii again misrepresents Famintsyn's familiarity with these matters. As will be shown below, the questions Vinogradskii investigated were treated by Famintsyn in his *Obmen Veshchestvo i Prevrashchenie Ehnergii v Rasteniiakh*, which was a broad compendium of information from botany, physiology, and chemistry reorganized into the first Russian monograph on plant physiology. In the sections that most closely related to Vinogradskii's microbiological interests, Famintsyn discussed in rich detail the work he considered most pertinent to those questions, including that of Nägeli and Pasteur. Zavarzin thinks that Vinogradskii's recollection of "meandering" was a misrepresentation of the laboratory dynamics. See G. A. Zavarzin, "Sergei Nikolaevich Vinogradskii (1856–1952)," *Khemosintez: K 100-letniiu otkrytiia S. N. Vinogradskim* (Moskva: Nauka, 1989), 5–21; See pp. 9–10.

of his emotional struggles as a Katzenjammer puzzle: "Was all his 'dissatisfaction and fatigue' worth the bother?" "Was it worthwhile to continue, and would it secure the honor of becoming a professor?"[68] Upon departing for Gorodok for the summer, he made the decision not to return to the laboratory. Again facing the dilemma characteristic of his generation and class, he felt caught between two worlds. The thought of the estate "vividly reminded [him] that [he] could have by rights settled there."[69] These feelings intensified once he arrived home and revisited the estate's garden, forest, and fields. Quickly drawn back into country work, he tried unsuccessfully to avoid dwelling on these career vacillations. The uncertainty, however, was "very much spoiling his existence—[he] felt like a failure, [and he] did not want to reconcile [him]self to those failures." He wanted to spend his days simply "living for [him]self, not philosophizing craftily (*ne mudrstvuia lukavo*)."[70]

In 1881, Vinogradskii made what he later called "the worst decision of his life" and returned to St. Petersburg University to continue his apprenticeship. Late in the fall, he recalled, "idleness sent me once again to the dreary, empty laboratory."[71] Perhaps the pressures of having children to raise—he now had two daughters, Zina and Tania—increased the pressure on him to "firmly establish a professor's career and to provide myself with the success of intensified work."[72] He later considered the next 2 years preparing for his Master's examination "superfluous" and "the most difficult in [his] life."[73] He "tormented [himself] with work, especially evening work, [and] fell into a neurasthenic condition, worrying [his] poor gentle mother and young children."[74] His work regimen during these 2 years became so intense and irritating that an impending sense of crisis loomed over his family.

In a usually gloomy mood, Vinogradskii conducted the investigation that would satisfy his master's degree requirements. His results were largely ignored by other scientists and, even to himself, seemed "a dead end and a waste of time."[75] The details of this investigation—the questions addressed, the organism studied, the laboratory methods applied, and the researchers and knowledge claims challenged—reveal, however, Vinogradskii's position on a variety of issues in plant physiology, providing our only window onto his values during this early, formative period. The investigation represents not only the store of knowledge and skills he acquired while at St. Petersburg University; it also provides us with a set of initial qualities against

[68] Vinogradskii, *Itogi*, 12.

[69] Ibid.

[70] Ibid., 12–13.

[71] Ibid., 13.

[72] Ibid.

[73] Ibid.

[74] Ibid.

[75] Vinogradskii, *Itogi*, 15.

which to view future changes in his laboratory practices and research concepts.[76] For example, his commitment to direct visual observations (and the techniques and apparatuses that accompanied it) during his apprenticeship would mature, developing through distinct stages during his career into a widely-adopted ecological method for soil microbiology in the 1930s. Vinogradskii's master's investigation reflects the lessons he learned from contemporary debates occurring in European science. Thus, to appreciate the significance of his first investigation we need to review the ideas, in which he was socialized.

The Cycle of Life in European Science

Vinogradskii had returned to Famintsyn in the 1880s, when Famintsyn was on the verge of setting plant physiology in a new direction. Outlining his grand vision in *The Exchange of Matter,* Famintsyn drew on Felix Hoppe-Seyler and Claude Bernard's ideas in general physiology, calling for the unification of physiology through investigations of the "two main vital functions: the acts of respiration and nutrition."[77] Famintsyn attached "primary meaning to the processes of nutrition" and, although he based his analyses on plant physiology, he believed they offered powerful analogies for the animal world.[78]

Through the study of nutrition, Famintsyn hoped to unite the entire organic world into one global economy of matter and energy. He envisioned a world in which "animals live on organic compounds prepared by plants or on animals which live on plant food; in other words they construct their bodies from the organic compounds prepared earlier by plants."[79] Plants themselves require "ready organic compounds for the construction of their cells, tissues and organs," which they obtain by transforming inorganic matter, "including a few mineral salts, water and carbon dioxide."[80] Since his earliest work in plant physiology, he had concluded that "plants" and "animals" were merely subjective morphological categories for very similar organisms. The proper way to understand the relationships between organisms was through the exchanges of matter and transformations of energy that occurred between them and their surrounding environments. Through the study of microorganism nutrition, he

---

[76] Zavarzin agrees: "Much of what [Vinogradskii] apprehended in this period made an impression on his entire life. Here was serious preparation as a systematist, and the incompatibility of systematics with the idea of pleiomorphism, an interest in the energetics of living beings, a command of microscope techniques, and a complete dismissal of any kind of professor's 'politics'." Zavarzin, "Sergei Nikolaevich Vinogradskii (1856–1952)," 9–10.

[77] Famintsyn, *Obmen Veshchestvo i Prevrashchenie Ehnergii v Rasteniiakh*, 13. See K. V. Manoilenko, "Rol' A. S. Famintsyn v Razvitii Ehvoliutsionnoi Fiziologii Rastenii," in Kursanov et al., *Andrei Sergeevich Famintsyn: Zhizn' i nauchnaia deiatel'nost'*, 131–149; see 140 for a short discussion of Hoppe-Seyler and Bernard's influence on Famintsyn.

[78] Famintsyn, *Obmen Veshchestvo i Prevrashchenie Ehnergii v Rasteniiakh*, 13.

[79] Ibid., 11.

[80] Ibid.

felt, the researcher would be investigating a crucial juncture in the circulation of matter and energy in nature—that between the inorganic and organic realms.

These were not new ideas in the 1880s.[81] Building on the late-eighteenth century ideas of Joseph Priestley (that plants restore the air used in animal respiration) and Antoine Lavoisier (who provided chemical interpretations of the reciprocal processes of respiration and vegetation) Jean Baptiste Dumas had proposed that plants possessed reduction apparatuses, and animals combustion apparatuses.[82] At the end of the eighteenth century, Lavoisier had introduced the quantitative method into chemistry and established that matter neither arises nor perishes, but that it exists in the same quantities throughout all of its alterations.[83] Other writers such as Priestley, Ingenhouss, Senibier, and Saussure had discovered the "principal laws of the transformation of matter in plants and animals, and thereby established the great doctrine of the circulation of matter in Nature, which portrays the organic and inorganic worlds in close reciprocal action."[84]

By extending the law of the preservation of energy, as formulated by Huygens and Leibniz, and Lavoisier's law of the conservation of matter, Robert Mayer applied their work to his physiological investigations in the mid-nineteenth century. Mayer's experiments on the relationship between heat and motion, through which he founded his law of the conservation of energy, demonstrated to him that in the course of vital processes forces were only transformed and never created.[85] It is probable that his ideas, as well as those of Hermann Helmholtz, strongly influenced biologists to consider the notions of the circulation of matter in nature and to study energy transformations in living organisms. The conservation of energy attracted little recognition between 1842 and 1860, but finally became established in the scientific literature because it could be profitably applied to develop novel investigations.[86]

When these ideas were flourishing, for example, in the work of Liebig and other physiologically minded chemists, Dumas became interested in elucidating the chemical transformations that took place within animals. Considering these

[81] Kremer, 454–455; Also, Hoppe-Seyler, "Vorwort," *Zeitschrift für physiologische Chemie*, Vol. 1 (1877), i–iii.

[82] Frederick Lawrence Holmes, *Claude Bernard and Animal Chemistry: The Emergence of a Scientist* (Cambridge: Harvard University Press, 1974), 15.

[83] This was considered by some to represent the conversion of chemistry into an exact science. Harald Höffding, *A History of Modern Philosophy: A Sketch of the History of Philosophy from the Close of the Renaissance to our Day*, trans. from the original German by B. E. Meyer (New York: Dover Publications, 1955), Vol. II, 493. On Lavoisier's contributions to chemistry, see J. R. Partington, "Lavoisier and the Foundation of Modern Chemistry," *A Short History of Chemistry* (New York: Harper and Brothers, 1960), Third Edition, 122–152.

[84] Höffding, *A History of Modern Philosophy*, 493.

[85] Ibid., 494. Robert Mayer (1814–1874) was a physician and physicist and published these ideas in his work entitled *Die organische Bewegung in ihrem Zusammenhange mit dem Stoffwechsel* in 1845.

[86] Ibid., 496–7. Mayer might have died in obscurity had it not been for the intervention of the English natural philosopher and microbiologist John Tyndall (1820–1894). Tyndall championed Mayer's work, and it was largely through his efforts that Mayer received research funding and eventually fame.

transformations to be combustions driven by the action of oxygen on organic compounds, he envisioned a program that would study the "complex partial oxidation reactions of organic compounds" using the new tools and achievements of organic chemists.[87] He presented these ideas in his last lecture at the Sorbonne entitled *Essay on the Chemical Statics of Organic Beings*.[88] The work of Dumas and Liebig during the 1840s, by redefining life in terms of chemical events, encouraged both animal and plant physiologists to transform their field into a strictly mechanical science.[89]

Dumas' vision of life captured Pasteur's imagination. Dumas's lectures, which Pasteur had attended, influenced Pasteur's understanding of the role of oxidation in fermentation, combustion, and putrefaction, and how those processes fueled the cycle of life. Pasteur would come to base his vision of a cycle of life on the tenet that "it is a law of the universe that all that has lived disappears."[90] He described the cycle of life as an "absolutely necessary" exchange of "mineral and gaseous substances," such as water vapor, carbonic gas (CO^2), ammonia gas, and nitrogen gas, from living beings back to the soils and atmosphere.[91] He thought of these substances as "simple and mobile [*voyageurs*] principles" that were moved to all locations on the planet by the movement of the atmosphere.[92] For him life drew on these materials in order to maintain its "indefinite perpetuity." What process, he asked, would cause living beings to relinquish their "simple principles?" He concluded that life was formed only where death and death's effect, decay [*la dissolution*] existed.[93]

From a much different background than Pasteur, the German botanist Ferdinand Cohn developed his own concept of the cycle of life. His mentor in botany, Christian G. Ehrenberg, who had dedicated his entire life to investigating the microscopic world, introduced Cohn to the study of the lowest animals and plants.[94] Cohn recounted that he was driven to study:

> these organisms that stand at the border between plants and animals" when it had come to light, "that the cell, in the clearest and most complete scientific investigations, was accessible

---

[87] Ibid., 20–22.

[88] Jean-Baptiste Dumas, *Essai de Statique Chimique des Êtres Organisés*, leçon le 20 août 1841, 2nd édition (Paris: Fortin, Masson, and Ce, Librairies, 1842).

[89] Höffding, *A History of Modern Philosophy*, 495; Geison, *The Private Science of Louis Pasteur*, 71–73, 88–89; and Partington, *A Short History of Chemistry*, 226–230.

[90] Pasteur, *Oeuvres*, vol. III, 84–85. Andrew Mendelsohn identifies and discusses this key passage in John Andrew Mendelsohn, *Cultures of Bacteriology: Formation and Transformation of a Science in France and Germany, 1870–1914* (Princeton University, Ph. D. dissertation, June 1996), 41–56, and esp. 45–46.

[91] Ibid.

[92] Ibid.

[93] Pasteur, *Oeuvres*, Vol. II, 648–653, esp. 653. This is a review written by Mr. Danicourt of an address present by Pasteur at the *Soirées scientifiques de la Sorbonne*, originally published in *Revue des cours scientifiques*, No. 18, February 1865; and cross-referenced under "*Vie*" in *Oeuvres*, Vol. VII, 657.

[94] Ibid. Christian G. Ehrenberg (1795–1876), Geheimer Medicinalrath and professor at the University of Berlin, at the time of instructing Cohn (in 1849) was using polarized light to determine the nature of microscopic objects. He did not arrive at any conclusions concerning their molecular structure, but this shows that Cohn may have been familiar with and possibly trained in the study of molecular structure. See Sachs, 354.

in those simplest, lowest, microscopic plants, in which, as single-celled beings, their entire development and complete life process took place in the same cell."[95]

Cohn's research on microbes took place in the context of the intellectual revolution in botany produced by Matthais Schleiden's cell theory and Hugo von Mohl's description of protoplasm in the plant cell.[96] Cohn's comparison of "the contractile contents of plant and animal cells," for example, "represented an important step toward the belief that the basic attributes of all life were to be sought in a single substance called protoplasm."[97] One aspect of the cell and protoplasm investigations included debates on the nature of fertilization. Schleiden's system of universal spore cells, which resulted in the understanding that plants reproduce both sexually and asexually, led Cohn to see the "study of developmental history as the key to all morphology, and the study of the cell's structure and life as key to plant physiology."[98] The quest to solve the mystery of fertilization, due to the crucial role it played in the life cycles of plants, was a prominent theme in botany during Cohn's university education.

In 1872, Cohn described his own view of the cycle of life in which "[t]he whole arrangement of nature" was based on the dissolution of dead organic bodies to provide the materials necessary for new life.[99] Due to the limited amount of material that could be molded into living beings, he conjectured that there must be a conversion of the same particles of material from dead bodies into living bodies in an "eternal circulation" [*ewigem kreislauf*].[100] For Cohn, bacteria were responsible for releasing the material bound up in each generation of plants and animals. By breaking down organic bodies, bacteria provide the "body material" needed for new life. He wrote that, "[s]ince bacteria cause dead bodies to come to the earth in rapid putrefaction, they alone cause the springing forth of new life, and therefore make the continuance of living creatures possible."[101]

---

[95] On November 14, 1897 Cohn recounted these feelings during a banquet in celebration of the 50 year anniversary of receiving his doctoral degree. See Cohn, "Rede bei dem Festessen der Universität," *Blätter*, 247.

[96] Geison conjectures that Cohn focused on the lowest plants, especially unicellular algae, because of Cohn's "conviction of the value of cellular studies and his belief that the best way to gain insight into the cellular processes of higher organisms was to begin by carefully studying the cellular processes of the simplest organisms." In his *Untersuchungen* and earlier work on *Protococcus pluvialis*, however, I find Cohn drawing on his knowledge of higher organisms to interpret his observations of lower organisms. See Geison, *Cohn*, 337.

[97] Ibid.

[98] John Farley, *Gametes and Spores: Ideas about Sexual Reproduction, 1750–1914* (Baltimore: The Johns Hopkins University Press, 1982), 82–85.

[99] Ferdinand Cohn, *Bacteria, the Smallest Living Beings* (Baltimore: Johns Hopkins University Press, 1939), 25; originally published as *Ueber Bacterien, die kleinsten lebenden Wesen*, in *Sammlung gemeinverständlicher wissenschaftlicher Vorträge*, Rud. Virchow and Fr. v. Holtzendorff, eds., (Berlin: C. B. Lüderitsche Verlagsbuchhandlung, 1872), ser. VII, vol. 165, 18.

[100] Ibid.

[101] Ibid.

The concepts of the cycle of life described above had much in common, yet their differences reveal the varied theoretical and methodological commitments of their authors. For example, Pasteur's concept revolved around the movement of simple principles in chemical processes while Cohn's centered on the life cycles of organisms. Pasteur envisioned a circulation of matter that included all the chemicals that provided living organisms their vital energy.[102] He viewed bacteria as chemical agents that operated through the process of combustion to release all chemical elements into the atmosphere. Life could then draw on this reserve as needed in order to maintain itself. He did not think the "circle of transformations" was complete until the microscopic beings had returned all organic material back into the atmosphere.[103] Cohn, on the other hand, emphasized the morphology and physiology of single celled organisms rather than the general flow of chemical elements (Pasteur's simple principles). For Cohn's "eternal circulation" of matter in nature to function, living beings would have to gain access somehow to the material stored in other organic beings.[104] The action of microbes, Cohn believed, would accomplish this exchange of matter. Vinogradskii would explore the significance of these subtle distinctions in his laboratory investigation of 1883.

---

[102] Pasteur's interest in investigating the chemical structure of life and nature can be traced back to his earlier notion of life in his crystal dissymmetry research. See Geison, *Pasteur*, 105.

[103] Pasteur, *Oeuvres*, Vol. II, 653. The *fleurs de vin* (flowers of wine) immediately makes water and carbonic acid, and vinegar is eventually produced. The vinegar, however, is still an organic material, and if the circle of transformations stops there, it will not be complete. These microscopic beings can continue their action and then, little by little, all the carbon and hydrogen of the vinegar changes into carbonic acid and water vapor, and thus all the organic material has passed back into the atmosphere.

[104] It is possible that this goes back to his support of the idea of alternating generations in microorganisms.

Chapter 2
The Exchange of Matter and the Transformation of Energy

Famintsyn's Approach to the Cycle of Life

As Famintsyn's apprentice, Vinogradskii's perspectives on questions of fungal nutrition agreed in large part with the views Famintsyn expressed in *The Exchange of Matter*. Famintsyn's 800 page monograph, which finally appeared in 1883, was a broad compendium of knowledge drawn from botany, physiology and chemistry. In concert with its sequel, a textbook entitled *Plant Physiology* (1887), it represented the founding of plant physiology as an independent discipline in Russia.[1] Although a quick comparison leads one to conclude that Vinogradskii's report mirrored the relevant sections of *The Exchange of Matter*, it was not, however, his investigation's sole source of inspiration.[2] Famintsyn's moody, inquisitive apprentice had also engaged the ideas of the broader European community of physiologists, microbiologists, and chemists through an exploration of a wide variety of sources. The most pertinent of these were Pasteur's publications on fermentation (which had made a deep impression on Vinogradskii) and Nägeli's work. It is most probable that Famintsyn, through his lectures and laboratory courses, had led Vinogradskii to study seriously Pasteur's and Nägeli's researches. For this reason I present their work as Vinogradskii first encountered it, filtered through Famintsyn's interpretation. The incongruities that remain, then, between Famintsyn's portrayal of certain questions in *The Exchange of Matter* and Vinogradskii's discussion of them in his presentation represent the latter's own contributions to the research.

Vinogradskii's investigation of the influence of nutrition and external conditions on the development of fungi cells was a continuation of Famintsyn's life work.

---

[1] See Strogonov, *Andrei Sergeevich Famintsyn, 1835–1918*, 70–71; and Kursanov et al., *Andrei Sergeevich Famintsyn: Zhizn' i nauchnaia deiatel'nost'*.

[2] I have considered that Vinogradskii introduced Famintsyn to the issues raised in the investigation, but all the evidence suggests that the flow of influence was towards Vinogradskii.

L. Ackert, *Sergei Vinogradskii and the Cycle of Life: From the Thermodynamics of Life to Ecological Microbiology, 1850-1950*, Archimedes 34,
DOI 10.1007/978-94-007-5198-9_2, © Springer Science+Business Media Dordrecht 2013

Vinogradskii's goal—to determine how fungi life cycles were related to the organic and mineral contents of their environment—was integral to Famintsyn's study of plant and fungal nutrition. Famintsyn considered plant nutrition to be only one dimension of the much grander process of the exchange of matter and the transformation of energy. This process not only provided the organizing theme of his book, but also, he believed, for all of nature. "The exchange of matter and the transformation of energy," he wrote, "are among the main functions of every living being; inseparably connected to them are all life functions, not only in the animal, but also in the plant organism."[3] He admitted that "the difference in organization between plants and animals is so apparent that at first glance it is difficult to show any analogy between their main life functions such as nutrition, respiration and reproduction, especially if considering higher representatives of the two kingdoms."[4] Yet the idea "that animals and plants share a common fundamental beginning of life and that a deeper and more attentive study of their most central vital functions will present much that is analogous" was gaining increasing respect in the physiological and biological sciences.[5]

Thus, for Famintsyn, the study of the simplest organisms—infusoria and algae—had "struck a decisive blow at the prevailing idea that a sharp border existed between animals and plants."[6] In 1860, he himself had contributed to this debate with his essay "On the Organisms at the Boundary of the Plant and Animal Kingdoms."[7] In this work, he supported his teacher Lev Tsenkovskii's view of life as a continuum from plants to animals, with microorganisms (infusoria) at the juncture between the two kingdoms. Tsenkovskii had explored this question from the perspective of systematics and morphology. Famintsyn, who had trained for 2 years in Europe with the mycologist Anton de Bary, introduced an approach based more on physiological, not morphological methods. Unger's discovery of *der Generationswechsel* (the alternation of generations—the successive appearance of forms with different modes of reproduction within a single species) in 1845 revealed the existence of what Famintsyn called transitional forms of simple organisms, which eventually destabilized most classification systems for microorganisms.[8] The crux of the problem for Famintsyn was that as long as "scientists were limited only to describing the

---

[3] Famintsyn, *Obmen Veshchestvo i Prevrashchenie Ehnergii v Rasteniiakh*, 10.

[4] Ibid.

[5] Ibid.

[6] Ibid.

[7] Andrei S. Famintsyn, "Organismy na granitse zhivotnago i rastitel'nago tsarstva," in *Sbornik Izdavaemyi Studentami Imeratorskago Peterburgskago Universiteta*, Vyp. 2. (S.-Peterburg: Tipografiia II-go Otd. Sob. E. I. B. Kantseliarii, 1860), 18–62.

[8] Famintsyn, "Organismy na granitsii zhivotnago i rastitel'nago tsarstva," 23. Famintsyn cites Unger's "A Plant at the Moment of Transforming into an Animal," as the first description of a simple organism (the single-celled alga *Vaucheria*) transitioning from an immobile to a mobile condition. See Famintsyn, *Obmen Veshchestvo i Prevrashchenie Ehnergii v Rasteniiakh*, 12. On the alternation of generations, see Lynn K. Nyhart, *Biology Takes Form: Animal Morphology and the German Universities, 1800–1900* (Chicago: The University of Chicago Press, 1995), 122.

shape and structure of microscopic beings, the attribution of the observed organism to one or another kingdom was determined by the subjective feelings of the observer."[9] The remedy came when scientists "began to study the developmental history of these simple representatives of life on the earth's surface."[10] Twenty years later, in 1880, the fire lit by Tsenkovskii and fanned by Anton de Bary was still burning in Famintsyn's mind.

Vinogradskii's interest in studying fungal development as a response to nutrition reflected Famintsyn's criticisms of most research on plant nutrition.[11] Famintsyn believed that the study of fungal nutrition had been hampered by focusing exclusively on "the nutrition of the simple representatives of the fungi class" (which included fermenting fungi), but he also thought the best work had been accomplished using those organisms.[12] His approach to fungal nutrition is reflected in the wording of his relevant chapters: chapter two, "Plant Nutrition by Organic Compounds" and chapter three, "The Synthesis of Organic Compounds in Plants." There, Famintsyn categorized fungi as "plants without chlorophyll," which allowed for investigating them using the same methods as "higher plants."[13]

This was more than a discussion of classification—the nutritional process revealed an organism's place in the flow of matter and energy in nature. Discussing the role of organic compounds in plant nutrition, Famintsyn explained that "[t]he necessary condition for the life of every living organism—both plants and animals—is the acquisition of food from without and its conversion into *organized* formations—cells and tissues; the life and growth of the organism are sustained by the exchange of matter with the surrounding environment."[14] The most graphic example of the conversion of nature's plastic materials through chemical transformations, were for him, the simple fungi. They grow, he explained, "in completely identical forms in the soil—which contains all their necessary organic compounds—and in artificial mixtures of mineral salts and sugar..... In the latter case they are compelled to manufacture inside themselves the albuminous bodies and fats necessary for the construction of fungal tissues."[15] Famintsyn criticized researchers working on fermenting fungi for wrongly concentrating on "changes in the substrate or the environment in which the fungi grow" while "almost completely ignoring the chemical metamorphoses which occur within the fungi cells."[16] Here he integrates the external and internal environments in an experimental model, one Vinogradskii would adopt and adapt.

[9] Famintsyn, *Obmen Veshchestvo i Prevrashchenie Ehnergii v Rasteniiakh*, 11.

[10] Ibid.

[11] This could also be related to Andrei Beketov's evolutionary views on the environment's influence on development. See Strogonov, 25–26; Todes, *Darwin Without Malthus*, 45–61.

[12] Famintsyn, *Obmen Veshchestvo i Prevrashchenie Ehnergii v Rasteniiakh*, 216, 375.

[13] Ibid., 140.

[14] Ibid., 141.

[15] Ibid., 142–143.

[16] Ibid., 216.

Physiological Debates on the Nature of Microorganisms

At the time of Vinogradskii's investigation, scientists categorized microorganisms, including *Mycoderma*, according to nearly as many classification systems as there were researchers. What names they used to distinguish their organisms reflected their stance on evolutionary and methodological issues. The choice between the terms *Mycoderma*, bacteria, *hefepilze*, and fungus, for example, reflected much about a researcher's position on the nature of microbial species. Were they determined by single, constant forms or life cycles; or were they more plastic, variable entities, directly responsive to changes in environmental conditions? A researcher's commitment to either of these positions influenced their experimental design and their interpretation of its results.

In his own investigation, Vinogradskii attempted to synthesize Famintsyn's reform agenda for plant physiology with the central tenets of Pasteurian microbiology and Ferdinand Cohn's taxonomic schema. Through these attempts he eventually developed an experimental regimen by which he could explore Famintsyn's grand vision of the transformation of matter and energy in nature in the laboratory. Vinogradskii's choice of experimental organism, one of at least 100,000 known fungi, reveals the concerns underlying the pedestrian conclusions of his 1883 report. He could have selected a multitude of organisms that Famintsyn considered either neglected or incorrectly investigated in plant physiology.[17] It is particularly significant that *Mycoderma* do not contain chlorophyll. Famintsyn had devoted the bulk of *The Exchange of Matter* to chlorophyllic plants—his own specialty had concerned photosynthesis—but had also discussed such non-chlorophyllic plants as fungi (including Vinogradskii's *Mycoderma*), molds, and bacteria. These 'colorless plants' had attracted the attention of plant physiologists because their inability to "nourish themselves on [inorganic] mineral compounds alone" meant that their surrounding environment must provide them organic compounds.[18]

Mycoderma thus satisfied the requirements essential to Famintsyn's agenda: it existed at the border between animals and plants, and its nutritional demands made it a viable organism for studying the exchange of matter and energy between the organic and inorganic realms. Vinogradskii had selected an organism his advisor placed at the cutting edge of plant physiology.

[17] Ibid. By choosing the "fermenting fungi" *Mycoderma vini Desm.*, Vinogradskii was, in part, satisfying Famintsyn's concern that, although a large number of investigations had been carried out on fungal nutrition, this work provided only a "scanty amount of information." It was also possible that Vinogradskii saw some practical application for studying this fermenting fungus, which also had been found to play a role in beet sugar production. Vinogradskii's family owned a presiding interest in a Kiev beet sugar plant.

[18] This investigation of this synthesis of organic matter, Famintsyn noted, had been initiated with marginal success in the three groups of microscopic plants: "all of which belong to the simple representatives of the class fungi: (1) the fermenting fungi, (2) several molds (Mucor, Penicillin and Aspergillus), and (3) the Schizomycetes (bacteria)." Ibid., 215–216, 375.

Vinogradskii's *Mycoderma vini* was suffering an identity crisis in the nineteenth century. Researchers' choices of species and genus names were telltale signs of their taxonomic preferences and the theoretical views that underlay them. Opposing schools of thought had produced various systems for classifying microorganisms. For example, Pasteur also used the name *Mycoderma vini*, while Ferdinand Cohn termed it *Saccharomyces mycoderma* and Nägeli included the organism in a broad class under the name *hefepilze* (yeast plants).[19] The variation in nomenclature reflected differences in these researchers' opinions about the role of microbes in fermentation and putrefaction (or decomposition).[20]

Cohn's understanding of bacteria ('bacteria' here included *Mycoderma vini*) and their role in putrefaction did not accord with what he called Pasteur's "paradoxical statement" that "putrefaction was a correlative phenomenon not of death, but of life."[21] In order to avoid the contradiction Cohn recommended "abandoning chemistry, which has studied the phenomenon of putrefaction only slightly" and being "content to wait until time gives us enlightenment, for now, we are restricted to establishing the biological relationship of the bacteria to putrefaction."[22] Although Famintsyn would have been uncomfortable eschewing chemistry in the 1880s, he would have agreed with Cohn's suggestion that fermenting fungi were able to assimilate the necessary materials from their environment and, through a material exchange, reshape (*umformen*) that material into the substance of their individual cells. Cohn even made the analogy to animals when he compared this process of "remodeling" to the process by which "animals transform the protein material of foodstuff into its flesh and blood during digestion."[23] Fermentation and putrefaction were, for Pasteur, processes of slow combustion (analogous to the faster "combustion by fire") that were caused by an intermediary agent.[24] Those agents were the "*infiniments petites*," and they played the "immense role" of "driving all other

[19] Other names included *Hermiscius cerevisea*, *Mucor mucedo*, and countless more.

[20] For a discussion of the wide variety of names for Mycoderma vini, see Oscar Brefeld, *Botanische Untersuchungen über Schimmelpilze* (Leipzig: Verlag von Arthur Felix, 1872), Heft 1, "*Mucor mucedo, Chaetocladium jones'ii, Piptocephalis freseniana. Zygomyceten*," 1–64; Max Resse, *Botanische Untersuchungen über die Alkoholgährungspilze* (Leipzig: Verlag von Arthur Felix, 1870), esp. 81–84. Famintsyn uses Reese's illustrations in his *Obmen Veshchestvo i Prevrashchenie Ehnergii v Rasteniiakh*.

[21] Ferdinand Cohn, "Untersuchungen über Bacterien," *Beiträge zur Biologie der Pflanzen* (Breslau: J. U. Kern's Verlag, 1875), Erster Band, Erstes Heft, (1870), 127–222, see 203–204. On Cohn see Gerald L. Geison, *Ferdinand Julius Cohn*, in the *Dictionary of Scientific Biography*, Vol. III, 336–341; and Pauline Cohn and Felix Rosen, *Ferdinand Cohn: Blätter der Erinnerung*, (Breslau: J. U. Kern's Verlag, 1901). Pauline Cohn (Ferdinand Cohn's wife) edited together a collection of Cohn's diaries and correspondence. Both of these biographical works provide extensive bibliographies.

[22] Cohn, "Untersuchungen über Bacterien," 204.

[23] Ibid.

[24] *Oeuvres de Pasteur* (Paris: Masson et Cie, Editeur's, 1939), ed. Pasteur Vallery-Radot, Vol. II, "Fermentations et generations dites Spontanées," 650.

beings to common ruin."[25] By using the term *Mycoderma vini*, then, Vinogradskii was associating his investigation with Pasteur's approach to fermentation.

In the Pasteurian spirit, Famintsyn had taught his students that fermentation was a process of fungal nutrition. It occurred in a two-stage process, as it did in other plants: "(1) the synthesis (although incomplete) of organic compounds, and (2) the consumption [of the organic compounds] for the building of organized formations."[26] He did not want to strictly demarcate these two phases of fungal nutrition, but did not doubt their existence, "since in fungus cultures in liquids that contain all the necessary organic compounds, the first phase of nutrition ends completely, and only then does the second appear."[27] By way of confirmation he cited Pasteur's demonstration "of the faster development of fungi in liquids that contain all the necessary organic compounds in the form of ready material for constructing its cells."[28] The problem with studying fermenting fungi, was that, even when using "irreproachably pure cultures," fermentation (stage two in fungal nutrition) significantly obscures those cultures, because it "halts the normal nutrition and growth of the fungi."[29] Vinogradskii's experimental protocols would need to consider these issues.

Even on the subject of alcohol fermentation, which had been "investigated incomparably better" than other nutritional processes, Famintsyn explained, the "most authoritative scholars" were in disagreement.[30] The debate concerned the role of fungi (or other microorganisms) in the fermentation process: Justus Liebig took them out of the process completely and "attributed fermentation to the molecular action of albuminous bodies … which decomposed the particles of other bodies into simpler particles." Maurice Traube and Felix Hoppe-Seyler gave them only a secondary role, suggesting that: "although the presence of simple organisms is necessary for fermentation, those organisms do not produce it directly; rather, they act "only as apparatuses that prepare the amorphous ferments" that cause fermentation.[31] Pasteur, Famintsyn wrote, viewed "fermentation as a function of the fungus, which is inseparably connected with its life; by his definition, fermentation is 'a phenomenon correlative to life.'"[32]

[25] Ibid., 653. One reason Famintsyn may have preferred Cohn's term, is that Pasteur supported Christian Ehrenberg's classification, which Famintsyn had dismissed as "unstable" in his "Organisms on the Boundary" (1860). See *Oeuvres*, Vol. II, 175, originally published as *Recherches sur la putréfaction, Comptes Rendus de l'Académie des Sciences*, séance du 29 juin 1863, LVI, 1189–1194.

[26] Famintsyn, *Obmen Veshchestvo i Prevrashchenie Ehnergii v Rasteniiakh*, 381.

[27] Famintsyn believed that "[o]ne of the next problems in the investigations of fermenting fungal nutrition will be the demarcation of these processes." Ibid.

[28] Ibid.

[29] Ibid.

[30] Ibid., 228.

[31] Ibid., 229.

[32] Ibid.

More recently, Nägeli, suggesting a new version of Liebig's mechanistic theory of fermentation, had returned the fungi to the process.[33] For Nägeli, the fermenting fungus' "living plasma" acts directly on its surrounding environment by transferring "the molecular vibrations that are inherent in its atoms and molecules."[34] These vibrations then produce "shakings" that cause sugar particles to decompose into smaller compounds.[35] For Famintsyn, then, despite the efforts of some of science's best minds, a final resolution of the question, "what is fermentation?" required "new, exact investigations."[36]

Vinogradskii was no doubt referring to the complexities of these debates when he compared his personal discomforts to the *Sturm und Drang* of the emerging field of microbiology. Entering the battle over these longstanding questions positioned him exactly where he wanted to be—amidst other creative innovators at the frontier of knowledge. Famintsyn offered some reassuring news: although the mysteries of fungal nutrition remained unsolved, that is, no one had "succeeded in defining exactly the reasons for the connection between the organisms' vital functions and the changes they conditioned in the surrounding environment"—someone, Pastuer, had pointed the way.[37] Pasteur's research on *Mycoderma vini* provided a good point of departure representing "the most thorough elaboration of the fermenting fungi's nutrition."[38] He had been the first to conduct exact and complete investigations of the products of alcohol fermentation, however Famintsyn noted, "[s]trictly speaking, cultures had almost never been created in which alcohol fermentation was completely eliminated."[39]

Fermentation, in Famintsyn's view, disrupted "normal nutrition" in fermenting fungi cultures and thus had to be prevented by controlling the availability of oxygen.[40] Moreover, unlike Pasteur, Famintsyn thought that alcohol fermentation should not be restricted to the fermenting fungi, for "it seemed to be a phenomenon common to all plants that are placed in conditions similar to those in which alcohol fermentation occurs."[41] According to Famintsyn's concept of the exchange of matter and the transformation of energy, the production of carbonic acid in developing (growing) plants (which he had observed in his own work on grapes) was "completely analogous" to the production of alcohol by fermenting fungi."[42] For him,

[33] Ibid. Famintsyn cites Nägeli, *Theorie der Gahrung* (1879), 174.

[34] Famintsyn, *Obmen Veshchestvo i Prevrashchenie Ehnergii v Rasteniiakh*, 229.

[35] Nägeli's idea differed from Liebig's view, in that where Liebig accepted that a simultaneous change occurs in the body producing the fermentation, Nägeli saw the plasma acting "completely mechanically, not undergoing any kind of alteration." Ibid.

[36] Ibid., 230.

[37] Ibid., 228.

[38] Ibid., 217–218.

[39] Ibid., 218.

[40] Ibid.

[41] Ibid., 225.

[42] Ibid.

"alcohol fermentation represents nothing more than a particular case of the exchange of gases [or matter]; it is an intramolecular, or internal, respiration."[43]

Vinogradskii, then, sought an "exact method" that would allow him to achieve Famintsyn's goal—to study intramolecular respiration during "normal nutrition."[44] With Famintsyn's guidance he found a method in Pasteur's notion of pure cultures. Famintsyn thought the most important condition for obtaining reliable results to clarify the chemical processes of fungal nutrition was "absolute culture purity."[45] Nutritional processes could be studied clearly, he wrote, only when "a single, selected, definitive form develops in an artificial mixture without the slightest trace of foreign organisms."[46] Surveying the literature, Famintsyn had found that most investigations fell into two classes: those in pure cultures based on a single "selected" fungal organism, "conducted with all appropriate precautions," and those of mixed cultures in which the target fungus was accompanied by various other fungi forms, which had accidentally fallen into the substrate.[47] His preference for "pure cultures of a single selected fungus" was not shared by all researchers. Some researchers, like Nägeli, held quite different views on the nature of microbial life—believing that their forms were not as steady or consistent as they were for Famintsyn and Vinogradskii.[48] Of the many works on fungal nutrition, only a very few had satisfied Famintsyn's methodological criteria, especially his insistence on strict conditions of culture purity.

An organism's ability to grow in particular environments revealed two processes central to late-nineteenth century physiology. How a plant developed—poorly, not at all, or luxuriantly—divulged the complex relationship that had developed between that plant and the energy and material reserves in nature's economy. These same characteristics of development also provided evidence about the internal workings of an organism's organs, tissues and cells. To investigate this relationship, Vinogradskii, unlike Famintsyn (who preferred Pasteur's double-necked retorts), chose Geissler chambers, which he had first encountered in Pasteur's *Etudes sur la Bière*.[49] These had all the advantages of Pasteur's retorts, but offered the advantage that they allowed him to directly observe changes in his organism's development as he adjusted the nutrient composition of the liquid substrates.[50]

---

[43] Ibid.

[44] At the time Vinogradskii began his investigation there were no chemical methods available to accomplish Famintsyn's objective of studying the chemical processes inside cells. Using the next best method, the prolonged, 'direct' observation of the organism's development, Vinogradskii surveyed his *Mycoderma* cultures for how efficiently or effectively they exchanged matter and transformed energy—that is, grew—in varied substrates.

[45] Ibid., 376.

[46] Ibid.

[47] Ibid.

[48] Ibid.

[49] Ibid. 377. Famintsyn was fascinated by "the peculiar simple, artificial solutions" used by investigators to provide most of what was known about fungal nutrition. The proper procedure to obtain a pure culture, in his opinion, was to grow the selected experimental fungus from a single spore in a sealed vessel, which contained a liquid substrate of some simple solution (both the vessel and substrate would have been sterilized by heating prior to adding the spore).

Vinogradskii's interests reached beyond glassware—he used the 'simple solution' known by then as 'Pasteur's Liquid.' This solution was "completely suitable for the normal development of fermenting fungi and for the synthesis of its main contents: albuminous bodies, cellulose and fat."[51] After Pasteur published his recipe for this solution, a frenzy of solutions began to appear in the chemical and botanical literature. Researchers, including Lothar Meyer, Nägeli, and Pasteur, debated the merits of these new laboratory tools offering simplified versions tailored for studying specific physiological processes.[52] In the process of these debates Famintsyn became highly critical of Nägeli's experiments, lamenting that they had not been conducted on "completely pure cultures." This compelled Famintsyn to regard all of Nägeli's experiments, and their results, with great skepticism.

Vinogradskii, using Famintsyn's methodological critique of Nägeli as a starting point for his research positioned himself among those botanists and physiologists for whom culture purity signified more than methodological rigor. When Vinogradskii confirmed Famintsyn's evaluation of Nägeli's findings regarding, for example, the necessity of potassium and alkaline-earth metals in fungal nutrition, he was worried about more than "the lack of confidence their scantiness and … the characteristics of their production methods" might have engendered. The care with which Vinogradskii designed and carried out his investigation, his tedious selection of nutritional substrates, vessels, and certain pure culture methods reflected his commitment to Famintsyn's scientific worldview. The exact method opened a portal into nature's grand circulation of matter and energy.

The debates described above provided the context for Vinogradskii's 1883 report. Three primary issues—the thermodynamic notion of a cycle of life, Famintsyn's translation of that concept into his transformation of energy and exchange of matter, and debates on the nature of microorganisms—set the stage for Vinogradskii's presentation. These are the issues, moreover, that connect the research he conducted during his apprenticeship with his later investigations in soil microbiology and ecology.

---

[50] See Pasteur, *Études sur la Bière*, 170.

[51] Ibid. Pasteur's liquid, contained in 100 cubic centimeters of water: 10 g of cane sugar, 0.1 g of ammonium tartrate, 0.07 g of ashes, obtained through the calcination of 1 g of dry yeast.

[52] Ibid., 379; Lothar Meyer offered the most popular "simple Pasteur solution," in which he "replaced the tartaric acid with nitric acid; the ammonia was introduced in the form of nitrates; and sugar remained the mixture's only organic compound." Nägeli conducted a long series of experiments testing the influence of a wide variety of these simple solutions, including Pasteur's and Meyer's, on fungi development. Famintsyn applauded Nägeli's extension of Pasteur's findings. Where Pasteur had "exactly determined the difference in fungal nutrition with and without the presence of free oxygen" using his single simple solution, Nägeli contributed "a lengthy series of interesting parallel experiments" on a much wider range of substrates. Famintsyn found it unfortunate that he could not accept Nägeli's interesting results "with unconditional reliability" and demanded "a careful verification." Famintsyn's main objection was that Nägeli, "in organizing his experiments proceeded from the undemonstrated position that all changes in the prepared mixtures were caused by the germs of simple organisms." Famintsyn applied the same reasoning to many of Nägeli's other experiments, which, due to the lack of culture purity, could not "provide exact information on the level of suitability of mixtures for the nutrition of each of the three groups of fungi."

The Intellectual Context for Vinogradskii's 1883 Report

On December 15, 1883, Vinogradskii synthesized his Master's investigation and delivered its results at the botanical section of the St. Petersburg Society of Naturalists.[53] The society often met in the university's auditorium, an amphitheater-shaped room with long tables and blackboards on a stage. On the table in front of him sat a pitcher of water and a glass, his demonstration materials and his lectern. This was not his first time in front of an audience—he had performed many solo piano recitals and concerts, so he may have been relaxed as he spoke to his audience. The title of his report was "On the influence of external conditions on the development of *Mycoderma vini.*" It concerned two problems: "(1) To find an exact method for studying the influence of external conditions on the development of lower fungi" and "(2) To investigate to what level the form of cells of some type of lower fungus remains constant in various conditions of nourishment."[54] The audience would have been familiar with his experimental organism, "the so-called *Mycoderma vini Desm.*"[55] (This microorganism had been under scrutiny by beer and wine makers, and later by microbiologists because of its role in fermentation.) Vinogradskii explained that he maintained his *Mycoderma* cultures in "nutritive liquids, the chemical make-up of which were precisely known."[56] These liquids differed from one another by only one component, while the remaining conditions of the culture were kept strictly uniform. Using "attentive" microscopical observation, "it was possible to link the observed characteristics of form and manner of growth in various nutritive liquids with the presence or absence of one kind of material in the liquid."[57]

Vinogradskii directed his audience's attention to the apparatus sitting on the table: an array of glass and rubber tubing, several retorts, a microscope, and several culture plates. His "method of observation," he explained, "is distinctive because the cultures were maintained in special apparatuses for microscopic culture, consisting of a Geissler chamber connected with gutta percha tubes to two glass vessels [which contained the *Mycoderma* cultures]."[58] The Geissler chamber was beneficial

[53] Sergei N. Vinogradskii, "O vliianii vneshnikh uslovii na razvitie *Mycoderma vini*," *Trudy Sankt-Peterburgskogo Obshchestva Estestvoispytatelei*, XVI, 2d ser. (1883), 132–135.

[54] Vinogradskii, "O vliianii vneshnikh uslovii na razvitie *Mycoderma vini*, 132. It is possible that in discussing the constancy of cell shape here, he is associating his investigation with debates on the nature of microscopic species. There is not enough evidence to make a strong case for this period of his life.

[55] As we shall see below this organism was known by several names at this time—Famintsyn referred to it *Saccharomyces Mycoderma*, Pasteur as *Mycoderma vini*, and Nägeli as *hefepilze*. It was more commonly known as wine yeast. [See Oscar Brefeld for an overview of this confusion. Also see F. Cohn's work.].

[56] Vinogradskii, "O vliianii vneshnikh uslovii na razvitie *Mycoderma vin*," 132.

[57] Ibid.

[58] Gutta Percha was a relatively new and increasingly popular item in chemical apparatuses during this period. It is "a rubber like gum produced from the latex of various SE Asian trees (esp. genera *Palaquium* and *Payena*) of the sapodilla family," see *Webster's New World Dictionary*, Second College Ed., Simon and Schuster, 1982.

in three important regards, he explained: it made it possible for him to observe the organisms for long periods of time, it allowed him the flexibility "to either supply fresh nutritive liquids continually or to alter the culture's conditions," and it "absolutely eliminated the danger of littering the cultures … with foreign organisms."[59] The Geissler chamber could also be housed under a microscope for viewing. Vinogradskii then proceeded to described his methods for initially sterilizing the cultures, and for maintaining their purity during the course of the experiment while introducing the organisms and the nutritive liquids.[60] He once again directed the attention of his witnesses to the demonstration table and the large covered plates (similar, no doubt, to Petri dishes) populated with healthy growths of *Mycoderma vini Desm*. In noted that order to guarantee "the complete uniformity of the material for observation," he had sowed his cultures with *Mycoderma vini* cells grown from a single cell and raised in "gelatinous cultures in covered dishes per Hansen's method."[61]

Vinogradskii then introduced his spectators to the specifics of his experiment. Using his apparatus, he had investigated two series of cultures: "in one, the organic materials of the nutritive liquids were altered and the minerals remained constant; in the other the nutritive liquids differed from one another by only one mineral component, and the organic materials were the same throughout."[62] For example, the first series began with an initial nutritive liquid containing glucose, peptone, citric acids and mineral materials. He altered this liquid by replacing peptone, first with ammonium tartrate, and then with leucine. He also replaced the sugar with alcohol, and raised and lowered the quantity of glucose and acidity. During each of these tests "the well-known characteristics of *Mycoderma*'s appearance, which are not observed in any other [organism], were successfully observed during protracted observations (of 18–20 days)."[63] *Mycoderma's* form did change, however, when Vinogradskii investigated "the influence of a large or small supply of oxygen on *Mycoderma* growth.[64] He found that the fungus demonstrated typical budding in a large supply of oxygen, but it formed mycelium[65] in an oxygen-deprived environment.

In a second series of cultures, the nutritive liquid common to all six apparatuses contained "glycerin, alcohol, asparagine, malic acid, ammonium phosphate, magnesium sulfate, and calcium acetate."[66] To each culture, Vinogradskii added one or

---

[59] Vinogradskii, "O vliianii vneshnikh uslovii na razvitie *Mycoderma vini*," 133.

[60] Ibid.

[61] Ibid. It is unclear from the report if Vinogradskii explained Hansen's method during his presentation.

[62] Ibid.

[63] Ibid.

[64] Ibid., 133–134. While this test may seem to reflect Pasteur's classic fermentation work, remember that Famintsyn had replicated Pasteur's work (and the related work of Boehm) on fermentation in grape cells. See above.

[65] The thallus, or vegetative part, of a fungus, made of a mass of threadlike tubes.

[66] Vinogradskii, "O vliianii vneshnikh uslovii na razvitie *Mycoderma vini*," 134.

two mineral salts: to the first, he added "potassium chloride, to the second—rubidium chloride, to the third—sodium chloride, to the fourth—potassium chloride and zinc acetate, to the fifth—potassium chloride and iron acetate, and finally, to the sixth—potassium chloride and arsenious acid."[67] He then surveyed these cultures for "the characteristics of form and growth of the *Mycoderma* cells," observing that these were "more or less sharply expressed" and the characteristics "[m]ost representative and consistent" were those in the culture [containing] zinc salts and potassium chloride."[68] Comparing this last culture with the previous one containing potassium chloride, he concluded that zinc salts produced a very "sharp" influence. *Mycoderma*'s cell growth was similarly influenced, he told his audience, by potassium chloride and sodium chloride.[69]

Vinogradskii concluded his exhibition with an attack on Nägeli's view of fungal nutrition. In two additional series of cultures, Vinogradskii tested the accuracy of Nägeli's earlier claims concerning especially the importance of mineral salts to fungus growth. For his first series, Vinogradskii grew six cultures of *Mycoderma* in Ehrlenmeyer retorts,[70] each containing a common nutritive solution of organic materials, phosphoric and sulfuric acid, and the salts of calcium and magnesium. He varied each culture by adding, in equivalent amounts, one of several alkaline earth metals (which he obtained from chlorine salts) in the following order: to the first, he added potassium; to the second, rubidium; to the third, cesium; to the fourth, sodium; to the fifth, lithium. The sixth culture, his control, remained "absolutely devoid of alkaline metals."[71]

Nägeli had claimed that "potassium was not absolutely necessary for fungal nutrition and could be replaced by rubidium or cesium," and that "sodium and lithium were not suitable for [fungal] nutrition."[72] Vinogradskii, however, had observed that *Mycoderma's* characteristic "luxurious film" had developed in only the cultures with potassium and rubidium. In the culture with sodium, moreover, he had observed "hardly any noticeable traces of growth" (the slight growth that did appear, he conjectured, was due, most likely, to "an insignificant trace of potassium chloride in the sodium chloride." He found that "not the slightest trace of growth" in the cultures containing cesium and Lithium. Nägeli was wrong.[73]

With perhaps a nod to Famintsyn and Krutitskii, who were waiting in the wings to present their own reports that day, Vinogradskii presented one final experiment to his audience. Nägeli had also claimed that fungi "necessarily required at least one kind … of alkaline earth metal."[74] To verify this assumption, Vinogradskii had

[67] Ibid.

[68] Ibid.

[69] Ibid.

[70] A container, generally of glass and with a long tube, in which substances are distilled.

[71] Vinogradskii, "O vliianii vneshnikh uslovii na razvitie *Mycoderma vini*," 134.

[72] Ibid., 134–135.

[73] Ibid., 135.

[74] Ibid.

prepared a series of four *Mycoderma* cultures using nutritive liquids containing equal amounts of organic material, phosphoric acid, and sodium chloride. To each culture, he added one of three alkaline earth metals salts in the following order: "to the first, magnesium sulfate; to the second, calcium sulfate; to the third, strontium sulfate; to the fourth, a control devoid of alkaline earth metals, [containing] only potassium sulfate."[75] His observation that *Mycoderma* cells grew only in the first culture (containing magnesium sulfate), in which they formed a "luxurious film," demonstrated, he explained, "that magnesium proved unconditionally necessary for *Mycoderma* nutrition" and "that the absence of calcium, as much as it is necessary for green plants, had no … favorable influence on *Mycoderma* development."[76] Once again, he concluded, his "observations did not support Nägeli's claims."[77] In this, his first public display, Vinogradskii firmly allied himself with Pasteur and Cohn, and the synthesizer, Famintsyn in the contested space of contemporary microbiological debates.

Vinogradskii's Style

Following Famintsyn's example, Vinogradskii synthesized varied methods and theoretical perspectives in his research, and this included the concept of the cycle of life that would inform his physiological and later ecological views of nature. Cohn and Pasteur, for example, represent a generation of investigators who, by thinking in terms of a 'grand cycle of life,' harmonized thermodynamic visions of nature with the new experimental methods of laboratory research—and they did so by focusing on the agency of microorganisms. Famintsyn's enduring interest in the physiology of single-celled beings had evolved into his theory of an exchange of matter and the transformation of energy. He indoctrinated his apprentice Vinogradskii with these ideas and taught him how to explore the vast and complex economy of nature—which traded on exchanges of chemical compounds—by using carefully designed and strictly controlled experiments. Famintsyn trained Vinogradskii to search for answers to life's mysteries in the nearly invisible world that flourished on the boundary of the organic and inorganic realms. These early lessons would serve Vinogradskii well during what proved a lengthy career of peering into this microscopic frontier. He nurtured the cycle of life world view through six decades of microbiological research. The cycle of life provided the underlying structure for his research from his first investigation in the 1880s, to his discovery of autotrophic life and chemosynthesis in the late 1880s, to his depiction of the nitrogen cycle at the turn of the century, and to his eventual redefinition of this concept as ecology in the 1920s–1940s.

---

[75] Ibid.

[76] Ibid.

[77] Ibid.

The immediate aftermath of Vinogradskii's December presentation to the Society of Naturalists, however, was far less grandiose: he left Famintsyn's laboratory. After completing his investigation and making his presentation, "the fumes began to clear and [he] recognized the danger [he] had brought to his family."[78] He decided "to immediately abandon the idea of a professor's career, and Petersburg with it."[79] Everything, he felt later, had gone so badly for him and he "had made a mess of [him]self and his work."[80]

Absent any career prospects, yet maintaining his love for science, he "lugged a microscope and boxes of instruments … to Crimea (where [he] spent the winter with Mama) and then to Gorodok."[81] It was not until the following November that he decided to continue his scientific research in Germany. He had come to realize that the time had passed in science when it was feasible to work as an amateur scientist in a domestic setting. "Here again," Vinogradskii grumbled, "I did not get sufficient advice about where to head."[82] He selected Anton De Bary's laboratory in Strassburg because Famintsyn and Voronin had previously been there. Vinogradskii, also, had "zealously" read De Bary's "excellent" works on mycology, which greatly inspired him.[83] Reflecting on his decision, he wondered whether "this new start would produce, like all the others, yet another failure?"[84]

[78] Vinogradskii, *Itogi*, 13. By "danger" Vinogradskii was imagining the dire situation his family would have faced had he died or become permanently injured from what he no doubt considered the health/life threatening conditions he had been working under.

[79] Ibid.

[80] Ibid., 13–14.

[81] Ibid., 14.

[82] Is seems unlikely that the idea for Vinogradskii to work in De Bary's laboratory had never been suggested by Famintsyn. Perhaps, however, Vinogradskii was too embarrassed at having quit the professor's track to ask for a letter of introduction.

[83] Ibid., 15.

[84] Ibid.

Part II
Experiment and Natural History

Chapter 3
The Laboratory is Nature: Investigating the Cycle of Life Under the Microscope

The tedium and torture that Vinogradskii associated with his apprenticeship in St. Petersburg did not long dissuade him from a scientific career. He spent the summer of 1885 recuperating and dabbling in scientific farming on his Kiev estate, still considering the same career alternatives that he had during his gymnasium days. The local botanical excursions he enjoyed with his mother and the experiments he conducted in his home laboratory attest to his enduring interest in botany, an interest that finally won out over a life as a gentleman farmer. That fall he departed for Strassburg and one of the foremost botanical teaching laboratories in Europe, where he would commit himself to a field of study that was itself just emerging from disciplinary chaos. During Vinogradskii's "Strassburg period"—the years from 1885 to 1888—he intensively investigated the physiology of several "peculiar" microscopic organisms.

Moving to Anton de Bary's laboratory in Strassburg marked Vinogradskii's entry into a new community of botanists and physiological chemists, the members of which he engaged on a variety of issues concerning technical approaches and theoretical problems. His approach to investigating the microorganisms he would ultimately classify as 'sulphur bacteria' and 'iron bacteria' reflected the influence of this community. He did not abandon, however, the perspectives he had learned under Famintsyn's mentorship. In part, Vinogradskii extended the line of research he had initiated in St. Petersburg by applying the same microcultural methods that he had used to explore the nutrition of *Mycoderma vini* to new experimental objects. The peculiar nutritional requirements of sulphur and iron bacteria led him to adjust his research objectives and methods. Moreover, this shift in approach and rhetorical strategy reveals Vinogradskii's commitment both to Famintsyn's thermodynamics approach to plant physiology (which focused on elucidating the exchange of matter and transformation of energy that occur during a plant's vital activities) and to a novel evolutionary perspective that was emerging during the 1880s in German botany.

Vinogradskii transferred his thermodynamic plant physiology to De Bary's laboratory at a time when botany was becoming ecological.[1] Envisioning nature as a

[1] See below and Part III for a discussion of the rise of ecology as a discipline and a way of thinking.

L. Ackert, *Sergei Vinogradskii and the Cycle of Life: From the Thermodynamics of Life to Ecological Microbiology, 1850-1950*, Archimedes 34,
DOI 10.1007/978-94-007-5198-9_3,

cycle of life in which matter and energy flowed between nature's inorganic reserves and its living organisms in thermodynamic exchanges controlled by the organisms' vital processes, Vinogradskii was fulfilling Famintsyn's research program. By exploring these vital processes—bacterial nutrition and respiration—as directly as possible through the creative use of biological, chemical, and physical techniques Vinogradskii and Famintsyn strove to reveal nature's cycle of life. Vinogradskii developed this cycle of life approach at the same time that a new Darwinist world-view was being espoused by a handful of German botanists who shared training in plant physiology, an interest in the problem of adaptation and natural selection, and experiences in foreign and exotic lands. These botanists were developing new approaches to plant anatomy, plant physiology, and plant geography that by the end of the nineteenth century would be identified as ecological.[2]

Vinogradskii's choice of experimental organism shaped the course of his investigations in Strassburg. His first organism had been rather mundane—*Mycoderma vini* grew in wine, beer or almost any other fermentable liquid; his new interest, *Beggiatoa* (sulphur bacteria), however, were the denizens of some of nature's most exotic places—swamps, marshes, bogs, and sulphur springs on steep Alpine slopes. *Beggiatoa*'s peculiar physiology, especially their nutritional demands, forced changes in Vinogradskii's research objectives and style. In response to these nutritional needs he reconfigured his technical repertoire—adding to the Geissler chambers and retorts used in his first investigation new variations of slide microcultures and, most important, "artificial environments" that enabled him to observe his wild (*sauvage*) organisms in what he thought might approximate their natural states. By correlating and comparing his observations across these cultures (back and forth between the laboratory and "free nature" as he called it) he reached novel conclusions about the role of sulphur bacteria in nature's economy that he considered more "natural" and less provisional than those of preceding investigators. His compulsion to substantiate his interpretation of laboratory experiments according to his natural history observations—a reflection of his thermodynamic vision of the cycle of life—inspired him to create versatile microcultural methods for his Strassburg research.

A Comfortable Internship: Anton de Bary's Laboratory at the University of Strassburg

Mentally depressed and physically exhausted after having successfully defended his Master's thesis in botany in the fall of 1884, Vinogradskii withdrew to the comfort of his family's estates. He spent the winter in Crimea with his mother, and the summer and fall until November in Gorodok basking in the southern sun surrounded by his

[2] Although a self-conscious discipline of ecology would not form until the early-twentieth century, the conceptual frameworks and methodologies that defined that discipline were already prevalent in late-nineteenth century botany and plant physiology.

extended family. As he emerged from this period of recuperation, however, there remained the same career questions that had haunted him earlier. Having abandoned the idea of an academic career without regret, he shifted his attention to practical subjects. He never seriously considered governmental or private service (viable career options available to him), preferring instead to settle in Gorodok, where he, with his botanical and microbiological knowledge, would occupy the vanguard of agriculture by applying scientific methods to that business.[3] Reminiscing late in life about the road not taken, he conjectured that he might well have become a leading agriculturist and forestry specialist.[4]

During these prolonged career vacillations his interest in botany never waned, as is clear from his organization of a home laboratory where he investigated cell morphology.[5] Entertaining "no illusions concerning the results of an amateur, domestic investigation," he decided to continue his formal science studies.[6] He later recalled that he was not as interested in an academic career, that is, in acquiring institutional status and teaching, as he was in conducting "laboratory investigations, interesting experiments in the Pasteurian spirit, and [making] 'discoveries'."[7]

The material resources essential for upgrading his scientific research from amateur to professional status could be acquired only via the professorial track. When he decided to pursue this academic route, however, he chose to study, not with Famintsyn in St. Petersburg, but with Anton de Bary, an influential botanist at the University of Strassburg.[8] At this critical juncture Vinogradskii again complained about a dearth of helpful advice. His choices, however, reflected the narrow selection of real opportunities available to him in science and the trends of his generation. In the second half of the nineteenth century, it was common for Russian students to continue their education in Western Europe, especially in Germany.[9] His specific

[3] He recalled that he could not pursue this option because "he was not able to set himself up." Though, if he had, he thought, he "would have probably remained in Gorodok [his] entire life, devoting [himself] completely to a country life and farm work, which had always attracted [him] …." This would have precluded any serious scientific career on his part." Vinogradskii, *Itogi*, 22.

[4] Ibid., 16.

[5] He does not specify what he investigated in this small laboratory, however, when he departed from the university in 1884 he took he microscope and glassware with him. At this time cell morphology was an expanding area of study that encompassed nearly every theoretical perspective and a wide range of techniques that could be preformed in a small home laboratory. Cell morphology would play a central role in his subsequent investigations at Strassburg and Zurich.

[6] Ibid., 16.

[7] Ibid., 14, 22.

[8] *Ibid.*, 13. The difficulties of his recent experiences in St. Petersburg, the freedom offered by his personal wealth, and a dislike for St. Petersburg's cold and damp climate certainly influenced him in making this decision, but it was also common for students (Russian, American, or other) at his stage of development to spend a couple of years training in Western Europe. There was a widespread perception within Russia's scientific society that any serious science investigator should study for a year or two in a Western university.

[9] Other options included working in with Simon Schwendener at the University of Berlin, Gottlieb Haberlandt in Tubingen, and Eduard Strasburger in Bonn, to name only a few. See Eugene Cittadino, *Nature as the Laboratory: Darwinian Plant Ecology in the German Empire*, 1880–1900 (Cambridge: Cambridge University Press, 1990) for a discussion of these schools.

choice of the University of Strassburg no doubt owed much to Famintsyn, who in the late 1850s had also studied with De Bary.[10] It was natural for Vinogrdskii to join De Bary's Strassburg laboratory. The botany departments at Strassburg and St. Petersburg were hardly foreign worlds; in many ways they spoke the same languages—literally, German and French, and metaphorically, by addressing similar theoretical topics and sharing practical objectives. Thus, De Bary would have been quite comfortable with Vinogradskii's thermodynamical approach to plant physio logy, his support of monomorphism and his application of direct methods to microscopical observation; and was perhaps eager to observe how his own ideas had been interpreted by the Russian school.[11]

Vinogradskii, moreover, was quite familiar and "greatly impressed" with De Bary's research.[12] Not only had he zealously read De Bary's mycological work when he was organizing his first investigation, but he had no doubt heard much about the famous German botanist from Famintsyn. Famintsyn's reminiscences of his rich intellectual experiences with De Bary may have alleviated Vinogradskii's anxiety about resuming a career in science. In any event, Vinogradskii found in Strassburg what he had sorely missed in St. Petersburg—a congenial scientific atmosphere. Characterized by a "pleasant note of simplicities and *gemütlichkeit,*" De Bary's laboratory fostered a harmonious union of professor and trainees and an "open exchange of ideas and demonstrations."[13] This setting, Vinogradskii found, facilitated pleasant and productive scientific work, free of the exhausting irritations he had experienced in Famintsyn's laboratory.[14] In contrast to "the sharp deterioration of psychological state" he had suffered during the torments of his Master's work, Vinogradskii rejoiced in the new working regimen with De Bary—and achieved virtuosity, not at a piano's keyboard or on the fields of his farm, but in the laboratory at the microscope.

---

[10] Famintsyn and De Bary shared an interest in investigating symbiotic relationships in fungi and lichens. De Bary had coined the term "symbiosis" in his *Die Erscheinung der Symbiose* (Strassburg, 1879). They showed each other a mutual respect, which would have enhanced De Bary's positive reception of Vinogradskii in 1885. Both Famintsyn and his close friend and colleague M. I. Voronin had, following the advice of their advisor Lev Semenovich Tsenkovskii, spent 1858 expanding their scientific horizons under De Bary's guiding charms. See B. P. Strogonov, *Andre Sergeevich Famintsyn, 1835–1918* (Moscow, Nauka, 1996), 22.

[11] Direct observations and the study of life cycles were central to De Bary's study of *Peronospora infestans* the parasite responsible for the 1861 potato blight. This research brought him wide recognition and involved him in the question of spontaneous generation. See Gloria Robinson, "Heinrich Anton De Bary," *Dictionary of Scientific Biography*, Vol. 1, 612–613; Vinogradskii, *Itogi*, 14.

[12] Vinogradskii, *Itogi*, 14.

[13] Ibid.

[14] Ibid.

A Brush with German Darwinism: Vinogradskii's Sulphur Spring Expeditions

Vinogradskii's research soon spilled out beyond the physical and intellectual confines of this congenial laboratory. Trained in natural history, he believed that laboratory experimentation in plant nutrition was a viable research objective as far as it explained the transfer of energy and matter as it occurred in nature—in the cycle of life. Conversely, such natural processes could be correctly investigated only through the rigorous application of meticulous techniques in controlled laboratory conditions. Vinogradskii was one of the very few botanists in the 1880s who combined natural history—in the spirit of Humboldtian botanical geography—with laboratory-based, experimental investigations. In Russia, as we have seen, he had imbibed a thermodynamic worldview in which there existed no meaningful boundary between nature and the laboratory. The intellectual conditions fostered by De Bary in his Strassburg laboratory allowed Vinogradskii to put this vision into practice.

Immediately after his arrival in November of 1885, Vinogradskii conducted an expedition to observe how his targeted organisms lived in their natural settings, and to collect samples of them for his experiments. Following the advice of Dr. Eduard Fischer, he visited four sulphur springs—three located near Lake Thun in the Bern Canton, Switzerland and one at Bad Langenbrück in Baden, Germany.[15] He found especially instructive the Rinderwald Spring (also called Fuchsweidli Bad) situated between the towns Frutigen and Adelboden along the Engstligen Valley, Adelboden Spring in Adelboden, and an unnamed spring near Leissigen.[16] Two of these springs were wild and difficult to approach; the others were either not used for medical purposes or were located in very primitive 'spas.'[17] The wild nature of these sites gave him the freedom to explore "through all sorts of drains, and water

[15] It is most probable that Fischer also directed Vinogradskii to Konrad Meyer-Ahrens, *Die Heilquellen und Kurorte der Schweiz: und einiger der Schweiz zunächst angrenzenden Gegenden der Nachbarstaaten* (Zurich: Drell, Füssli, and Co., 1867), 2nd edition. In this extensive classification of mineral springs and health resorts in Switzerland Meyer-Ahrens described all of the springs that Vinogradskii visited: Adelboden, Rinderwald, and unnamed spring near Leissigen. It is quite clear that Vinogradskii referenced Meyer-Ahrens book; the description of the Leissigen source follows Meyer-Ahrens verbatim. Vinogradskii cites him in another place in "Ueber Schwefelbacterien." Meyer-Ahrens lists Bad Langenbrück in the Canton Basel, Switzerland; see his *Die Heilquellen und Kurorte der Schweiz,* 659–660. This cite was popular with the chemists, including Robert Bunsen and his student Lothar Meyer; The physician Charles Müller, who Vinogradskii cites, also tested these waters; see Meyer-Ahrens, see *Die Heilquellen und Kurorte der Schweiz,* 226.

[16] Winogradsky, "Ueber Schwefelbakterien," 530–531.

[17] Ibid., 531. He does not mention for what purposes the other two were used.

pipes" from the very openings of the springs along its longest stretches.[18] These were not the more common hot springs associated with spas, with water temperatures of 5–8°C.

At these springs he found the object of his search—luxuriant growths of snow white *Beggiatoa*-velvet. *Beggiatoa* is a genus of filamental bacteria that grow in the presence of decaying matter and abound in sulphur-rich waters. For the most part undetectable, in optimal conditions such as in sulphur springs, *Beggiatoa* can form huge blooms of snow-white, globular masses.[19] The Leissigen spring protruded from a steep mountain slope where water flowed out through a wide clay pipe, past a marshy patch and onto a stony bed. Here, he noticed, the *Beggiatoa*-velvet covered the clay pipe (wherever the water washed against it) and the stones, wood, and fallen leaves along the bottom of the stream about 10 m from the opening. To convince himself (although it was hardly necessary, he wrote) that the velvet was "not some kind of slimy covering, possibly consisting of dead filaments," he removed the *Beggiatoa* covering from the clay pipe as far as he could reach using his knife and hand.[20] When he returned after 10 days the "pipe was already clothed again just as thickly with a *Beggiatoa* cover."[21]

Vinogradskii's preoccupation with the nutritional relationship between organisms and their environment guided his observations. The Fuchsweidli spring emerged from a small pool 1 m deep, flowed out onto a bed of covered with rounded stone, gravel, and sand, and traveled 10 m into a large brook. The *Beggiatoa* did not grow in the pool itself, but an unbroken snow-white cover did form over the stone bed. Where the spring emptied into the brook (which contained much more water than the spring), "the *Beggiatoa*-cover ended as if cut off." At the Adelboden spring the *Beggiatoa*-cover was noticeable only at the spring's very opening. These observations led Vinogradskii to correlate the formation of luxurious *Beggiatoa* growths with the presence of free hydrogen sulfide in the spring water. At each spring, he noticed, when the strong scent of hydrogen sulfide disappeared, so, too, did the *Beggiatoa* growths. This indicated "with utmost certainty that the hydrogen sulfide exerts a favorable influence on the nutrition of these organisms."[22] He was most surprised that *Beggiatoa*, not containing chlorophyll, could develop in conditions where neither other bacteria nor green algae lived.

On what did they subsist? Because organic matter existed in such minute amounts in the sulphur springs he surmised that they must need only a very little.

---

[18] Ibid. Vinogradskii knew that these sulphur springs belonged to the larger sulphur spring group, which arose from the gypsum rich Jurassic limestone of the Stockhorn and Nieson network of mountain ranges; see Ibid., 531.

[19] Today *Beggiatoa* are one of the primary organisms used in microbial ecology for determining the health of ecosystems. Because it is difficult to raise this organism in artificial cultures laboratory investigators must still find natural sources.

[20] Ibid., 557.

[21] Ibid., 557–558.

[22] Ibid., 531.

It was probable, then, that even though this organic material was available to the *Beggiatoa* in a highly diluted state it sufficed for their needs. He imagined that they would have access to a constant supply of this organic material because the water flowing into their surrounding environment continuously refreshed it.[23] Yet to be convinced of his conclusion he required more than observations made "in free nature" *(in freier Natur)*. His descriptions of these sulphur springs demonstrate that he was primarily concerned with nutritional processes. The question most important to him—one that had not been answered anywhere in the literature—was: what organic material (nutrients) existed in the sulphur water?[24] As we shall see, this question set the initial direction for his Strassburg investigations of sulphur bacteria and other associated species.

The inspiration for Vinogradskii's sulphur spring expeditions came from several directions. He could have fulfilled the requirements of the investigation De Bary had recommended—to investigate whether certain species of microscopic fungi maintained constant life cycles in varied environments—by simply using samples of *Beggiatoa* already available in the Strassburg laboratories or local collections.[25] In fact, De Bary himself had investigated these organisms and most likely already had several species in his possession. Vinogradskii was also aware that *Beggiatoa* grew wherever organic matter was decaying in slow moving or still waters and he had earlier collected samples in the University of Strassburg's botanical garden. That he made these difficult trips to such inhospitable sites reflects his commitment to his own general approach to the study of nature. Only in free, wild nature—undisturbed by human activity—did organisms exist in a natural relationship to material resources. Because this relationship represented the true role of an organism in nature, it was his observations at these wild sites, and not at the spas, that furnished the strongest evidence for his published claims.

Spa life in the 1880s had taken on new dimensions—becoming an activity not only for the elite, but also for the middle classes. This rise in popularity increased the awareness of sanitation experts about the possible health benefits and risks of the minerals and organic life prevalent in the spas. Natural historians had also maintained a long-standing interest in these mysterious places. Vinogradskii had familiarized himself with this scientific literature and had probably 'taken the waters' at some time.[26] However, although there were plenty of easily accessible hot sulphur springs around Strassburg, he planned his expedition around other destinations. He was reluctant, perhaps, to collect his samples amidst the thermal spas' patrons, who would have been soaking up the sulphur air, having the black mud applied to their

[23] Winogradsky, "Ueber Schwefelbacterien," 557–558.

[24] The best source he had located was an analysis of the water at the Weilbach spring in Fresenius, *Jahrbucher des vereins fur Naturkunde in Herzogthum Nassau*, Heft. XI, 1854.

[25] It was common practice in this community to share samples, even across disciplinary lines and with challengers to one's claims.

[26] There is evidence that the Vinogradskii family enjoyed the spas. Vinogradskii often took his family to the Evian-aux-Bains and the Aix-aux-Bains when he lived in Brie-Comte-Robert near Paris in the 1920s–1940s.

bodies, and drinking the mineral waters.[27] For medical researchers and naturalists, it would have been quite acceptable to fill their numerous large flasks with the waters and especially the bright white *Beggiatoa* globs that filled the springs. These patrons—primarily middle-class tourists and locals—would not be surprised to witness scientists collecting samples. Sanitation experts, botanists, and chemists had visited often over the past decades to test the waters for diseases and to explore some of nature's most peculiar spaces. As a naturalist searching for the secrets of free nature, however, Vinogradskii's theoretical commitments demanded that he seek sulphur springs in the wild, far from any medical institutions and human activity. Only here in savage nature did organisms play out their specific roles in the economy of nature. In the wild springs they existed relatively undisturbed by influences that would have altered their natural state. Vinogradskii's choices reflected his devotion to the venerable traditions of natural historians, a devotion tempered by his enthusiasm for experimental biology and a rising movement in evolutionary botany.

The novelty of Vinogradskii's approach to plant physiology emerges when viewed against the background of developments in German botany during the 1880s. Eugene Cittadino has identified a generation of German botanists who were part of an entrenched laboratory tradition, practiced in the tenets of Darwinian evolution, and had made significant contributions to Darwinian biology.[28] The feature of their training most important for understanding their development of an ecological botany was their experience in the exotic landscapes found in the Reich's new African and South American territories. Foreign travels provided them with a much different perspective than that which had emerged from their previous research on European flora. Cittadino's Darwinian botanists, including Gottlieb Haberlandt, Georg Volkens, Ernst Stahl, and A. F. W. Schimper,

[27] On spa culture see: Helena Wadley Lepovitz, "Pilgrims, Patients, and Painters: The Formation of a Tourist Culture in Bavaria" *Historical Reflections*, Vol. 18, No. 1, 121–145, esp. 123, 126, 144–145; The "pilgrims and patients" that Lepovitz describes as journeying "to natural curative locations set amidst the ethnic vistas celebrated by Bavarian artists" shared their sites with botanists such as Cohn and Vinogradskii who also conducted repeated pilgrimages. Where Lepovitz's pilgrims and patients were helping to shape Bavaria's tourist culture, the nineteenth century naturalists were shaping European scientific culture by engaging certain networks, investigatory methodologies and over arching theoretical systems. Douglas Peter Mackaman, "The Landscape of a Ville d'Eau: Public Space and Social Practice at the Spas of France, 1850–1890," *Proceedings of the Annual Meeting of the Western Society for French History*, Vol. 20, 1993, 281–291; Mackaman writes that "Hesitant and uninitiated travelers consulted one of the countless spa guide books written by prolific publicists like Louis Berthet. These books beyond their vast amount of technical information, attempted to give their readers something of the traveler's sensibility and experience. One learned much about otherwise foreign "topographies" by reading spa guides, as most books lavished a great deal of prose on carefully crafted descriptions of terrain, geology, meteorology, flora, and fauna. In discussing natural science and local history, authors typically conferred a combination of monumentality and romanticism on their subject." Mackaman, "The Landscape of a Ville d'Eau: Public Space and Social Practice at the Spas of France, 1850–1890," 283.

[28] Cittadino, *Nature as the Laboratory*, 1–2.

organized studies of plant adaptation from 1880 to 1900, seeking to extend the scope of botany to encompass the relationship of individual plants to their environment.[29] Swept along with the Reich's colonial expansion, they traveled to tropical lands to investigate how organic form and function related to environmental factors. There they took advantage of the extreme environmental conditions of heat and moisture unfamiliar to Europeans.[30]

In some ways, Vinogradskii fits this profile quite well—he, too, trained in botany and was anchored in a well-established laboratory tradition during the post-Darwinian period. The differences that set him apart from this cohort, however, are significant and telling. He came from a very different geographical setting with a distinct complement of flora and fauna—for him, Strassburg was already a foreign land. He needed to go no further than the poisonous waters of sulphur springs located on steep Alpine slopes to find the extreme conditions and exoticism so sought by his peers.[31]

As Cittadino has noted, these botanists belonged to the first generation to come of age in the "Darwinian universe."[32] Although Vinogradskii, too, studied the natural sciences during this post-Darwinian period, he never explicitly identified himself as a Darwinian or discussed evolutionary theory.[33] By training in Russia and Germany "when Darwinism was enjoying its greatest popularity," however, he could not have avoided considering the relationship, so important to his German laboratory peers, between natural selection and the complex adaptations of plants to environmental factors.[34] If he were at all amenable to the German botanists' support of Darwinian evolution by natural selection, his support would have been moderated by his

---

[29] Ibid., 4.

[30] Ibid., 5–6.

[31] Although other scientists had visited these same or similar sulphur springs, and had also noted their peculiarity, Vinogradskii interpreted these sites very differently. Below we will see that Ferdinand Cohn, for example, who, like Vinogradskii also viewed the role of bacteria in nature's economy from a "biological" perspective, had not conducted his collecting in the same way as had Vinogradskii.

[32] Ibid., 4. I agree with Cittadino's portrayal, but am reluctant to accept his stress on 'younger men' and 'new generation,' for the worst of its inherent Kuhnian implications. This history is complicated also by the fact that German transformism itself has a much longer history that predates Darwin's work. See Pauline M. H. Mazumdar, *Species and Specificity: An Interpretation of the History of Immunology* (Cambridge: Cambridge University Press, 1995), 34. She cites Philip R Sloan, "Darwin, Vital Matter, and the Transformation of Species," *Journal of the History of Biology*, Vol. 19, 1986, 369–445.

[33] I have not found a single reference to Darwin's work in any of Vinogradskii's writing. In his sulphur bacteria research he does refer to competition between species for resources, but the idea of competition for resources has a long history that predates Darwin's evolution by natural selection.

[34] On the introduction of Darwinian evolution into German botany see Cittadino, *Nature as the Laboratory*, 2, 4; and on the reception of Darwinism in Russia see Daniel P. Todes, *Darwin Without Malthus: The Struggle for Existence in Russian Evolutionary Thought* (New York: Oxford University Press, 1989), 23.

education in Russian evolutionary theory. Like many Russian biologists, for example, he may well have found fault with the cold Malthusianism inherent in Darwin's central evolutionary mechanism—a criticism emphasized by his St. Petersburg mentors Andrei Beketov and Famintsyn. Vinogradskii's grounding in Pasteurian microbiology also may have inclined him towards an anti-evolutionary perspective, or at least confused the issue to the extent that he felt uncomfortable with the topic. For whatever reasons, he did not set his studies in an obvious evolutionary context.

Vinogradskii diverged most significantly from the German botanists by his extensive training in and commitment to plant physiology. Stahl and Schimper distinguished their new "plant biology"—the investigation of how environmental factors such as moisture, light, and temperature influenced plant anatomy—from plant physiology, which they considered less interpretive and too specialized. In sharp contrast, Vinogradskii's research on microorganisms, and specifically on their vital processes of nutrition and respiration, demanded the specialized technical manipulations that the German botanists had relinquished when nature became their laboratory. This early divergence between Vinogradskii and the German Darwinian botanists would expand over the next four decades and into the twentieth century. At that time, the efforts of these plant adaptationists would influence the founding of the new science of plant ecology. Vinogradskii's perspective would also contribute to the founding of new ecological disciplines—in soil science and microbiology.

Species Constancy: Vinogradskii's Physiological Interpretation of the Monomorphism-Pleiomorphism Debate

Debates on the nature of speciation, one of De Bary's driving interests, influenced Vinogradskii's research agenda in Strassburg. In the late fall of 1885, he returned to De Bary's laboratory, arms laden with collection bottles and prepared to begin his new assignment. He was instructed to investigate recent claims challenging De Bary's assumption that a class of microscopic fungi (including *Beggiatoa*) maintained constant life cycles. If De Bary's ideas could be corroborated, the next task would be to establish a new classification system for these organisms. Thus, Vinogradskii's assigned project was central to De Bary's own interest in investigating the stability of microorganism species. Yet, as we have seen above, Vinogradskii was intimately familiar with this question, and no doubt consented to pursue it because it formed the core issue in a long-standing debate about the fundamental nature of microbes.

The microbiologists who participated in this dispute are as difficult to classify, perhaps, as the microorganisms they studied, but they generally can be divided into two groups—the monomorphists and pleiomorphists. Monomorphists saw stability in nature and identified species of microorganisms by the mature stage of their relatively regular and stable life cycle. For monomorphists, microbial species could thus be classified much like the higher plants and animals. Pleiomorphists, on the other hand, explained the extraordinary variation in shape observed in the microbial realm by the extreme plasticity of a few real species. Pleiomorphists believed that these few species (some proposed that only two existed) changed shape according to their conditions of existence; consequently, any microbial taxonomy was at best provisional.

Vinogradskii's own views and practices intersected these debates only tangentially.[35] Rhetorically he supported a staunch monomorphist position—advocating for the views of Cohn and De Bary against those of the pleiomorphists, as expressed primarily by Ray Lankester, Eugenius Warming, and Wilhelm Zopf. Whether it was solely due to De Bary's encouragement or whether the nature of the debate itself suggested it, Vinogradskii focused his attacks primarily on Zopf, one of the most outspoken pleiomorphists and an outspoken critic of De Bary.[36]

The willingness on the part of De Bary and Vinogradskii to accept that microscopic fungi did at times exhibit pleiomorphic characteristics raises an important question about their reasons for investigating these issues. To address this issue and, thus, to understand Vinogradskii's reasons for collecting bottles of sulphur spring water, we need to briefly explore the monomorphism-pleiomorphism debate and Cohn's attempts to establish a bacterial classification system.[37] The mysteries surrounding microbial shapes and physiological processes that intrigued Vinogradskii

[35] With few exceptions, previous portrayals of Vinogradskii's discovery of new physiological types and later chemosynthesis have relied too heavily on his own portrayal of the events, especially as related in his autobiographical remarks. Most historians have drawn primarily on *Microbiologiia Pochvy* (the Russian translation of the French compendium of his life's work), or his own account of this period in other autobiographical writings. Few, it seems have spent time with the original German and French published reports which contain a wealth of information, relative both to his scientific efforts and personal history. It is difficult to tease out the multiple influences on his research during this period. On the monomorphism-pleiomorphism debate see Penn and Dworkin (1976), Amsterdamska (1988), and Vinogradskii (1937); James Strick, *Sparks of Life: Darwinism and the Victorian Debates over Spontaneous Generation* (Cambridge, MA: Harvard University Press, 2000), esp. 123–128; and Mazumdar, *Species and Specificity: An Interpretation of the History of Immunology*, 46–59.

[36] Famintsyn conducted novel researches into the nature of photosynthesis, and viewed the interaction between organisms and their environment as a dynamic, not static process. His ideas were very influential on Vinogradskii's research.

[37] Vinogradskii did not claim that a pleiomorphic species could not exist; however, he would accept such a finding if it some future work could demonstrate it, if it used his prescribed method of prolonged direct observation of life cycles. In his own experiments, however, he had found no evidence in support of any such 'natural law' for bacteria. Although this debate did play a part in Vinogradskii's research at Strassburg (and sporadically throughout his career), and though it spread throughout the scientific community, especially when it involved mycology (the study of fungi), this is not the place for an extensive discussion of its details. He was aware of the debate and even engaged it quite openly. Explaining the sulphur bacteria research that Vinogradskii conducted in Strassburg solely in terms of the species constancy question restricts the significance of his broader theoretical commitments to plant physiology. Although he did believe that by focusing on physiological traits he had solved the riddle of some bacterial genetic relationships—that is, he had proven pleiomorphic taxonomies incorrect—he was more concerned with trying to understand their roles in the economy of nature.

were generated within this intellectual context.[38] As early as the late-eighteenth century botanists had been intrigued by microscopic organisms that were associated with a variety of natural elements, such as sulphur and iron. One such organism, *Beggiatoa*, had become central to the longstanding debate between monomorphists and pleiomorphists over species constancy. As monomorphists, De Bary, Vinogradskii, and Cohn were committed to the notion of stable microbial species.[39] De Bary and Vinogradskii, however, also allowed that "species … may appear in very unlike forms even in the same [periods] of development" and that pleiomorphic species differed from monomorphic ones by "their greater amount of differentiation during the course of development." In their investigations they adopted the bacterial taxonomy (which included microscopic fungi) published by Cohn between 1872 and 1875. By 1885, when Vinogradskii entered the fray, the literature on bacteria had increased dramatically, including the appearance of several monographs in the first half of the 1880s. Drawing widely on this extensive literature, Cohn established a comprehensive bacterial taxonomy that, although it may not have completely satisfied anyone including the author himself, earned widespread recognition and became the new point of departure for subsequent classification systems.[40] Vinogradskii lent his support to Cohn's bacterial taxonomy not only because it was the most precise description of the natural organization of microorganisms, but also because of the classificatory role Cohn ultimately assigned to physiology.

---

[38] There may also have been some distant relation between Vinogradskii's interest here and Nägeli's proposed iconoplasm. Nägeli described iconoplasm as "a filament of "micelles," which crystallized out of "the primitive albuminoid matter in the ooze where life begins." See Charles Coulston Gillespie, *The Edge of Objectivity: An Essay in the History of Scientific Ideas* (Princeton: Princeton University Press, 1960), 323.

[39] De Bary, *Lectures on Bacteria*, 25–26. The perspective espoused by Zavarzin and others that Vinogradskii pursued his research in Strasbourg and Zurich as a conscious plan to prove the fallacy of the pleiomorphic view of microbial systematics, and that this research program stemmed logically from his training in St. Petersburg, is not supported by the available evidence. Zavarzin writes that "A botanist by education, Winogradsky had strong objections to the theory of pleiomorphism in bacteriology and the realization of the rigidity of bacterial form and function came to be one of the leading ideas spanning all his research work." (G. A. Zavarzin, "Sergei N. Winogradsky and the Discovery of Chemosynthesis" in Hans G. Schlegel and Botho Bowein, eds., *Autotrophic Bacteria* (Madison: Science Tech Publishers, 1989); This book is a collection of reviews based on the symposium "Lithoautotrophy, a centenary meeting in memory of S.N. Winogradsky" held in Gottingen, August 23–28, 1987). There is, however, an interesting coincidence in this debate. Zopf (also a botanist) published his *Zur Morphologie des Saltpilze* in 1882 (the same year that Tsenkovskii published *Microorganismy*) in which he described his theory of 'the changeability of forms in response to various substrates." Here Zopf recognized his Russian supporters, Tsenkovskii, Gobi and Kostychev (the latter two had published a Russian translation of Zopf's monograph. Russia's founding microbiologists (primarily botanists at this time, excluding Metchnikov) were divided on this question of microbial species constancy. Pleiomorphists included among their numbers Tsenkovskii, Gobi, Kostychev, Ivanov, Metchnikov, and Kholodnyi. The monomorphists were Famintsyn and Voronin.

[40] Ferdinand Cohn, "Untersuchungen über Bacterien," *Beiträge zur Biologie der Pflanzen* (Breslau: J.U. Kern's Verlag, Max Muller, 1875), Vol. 1, No. 2, 1872, 127–222; Ibid., "Untersuchungen über Bacterien II," *Idem.*, Vol. 1, No. III, 1875, 141–204.

Recognizing the futility of trying to organize the microbial realm into a system that represented the natural, genetic relationships of bacterial species, Cohn opted instead for an artificial arrangement. In the research that led to his classification system, he first endeavored "to produce an independent opinion (*Urtheil*) on the biological relationships [of bacteria] as well as on the division of [their] species."[41] In the face of the extraordinary taxonomic chaos he encountered in the literature, he wanted to determine which organisms belonged to the category bacteria and which genera (*Gattungen*) and types (*Arten*) could be distinguished among them. The difficulties in establishing a classification system analogous to those for higher plants and animals were mostly related to the extremely small size of the bacteria. Not only was it impossible to differentiate between the internal contents of the bacteria, Cohn told his readers, but it was even difficult at times to discern if the organisms consisted of a single cell or a chain of two cells. It was also impossible to isolate single bacteria and to observe them under varied conditions for long periods of time, and to be certain when sowing mass cultures of bacteria that only the one species being investigated had been introduced into the culture. These difficulties were significant barriers to classifying, he believed, only if one intended to separate bacteria into "natural genera." Attempts to establish such natural bacterial genera along the lines of those for the higher organisms had failed because they were founded, not on reproductive characteristics, but only on "vegetative cell configurations" (i.e., the shapes of their colonies).[42] Admitting that "for now we possess no method to delimit the age of bacteria, their developmental state, varieties and kinds," Cohn was forced to adopt a substitute. To investigate the microbial world, researchers were compelled, he felt, to borrow a method still popular with mycologists and paleontologists—a method by which a genus was defined by its most striking features, with each small deviation from that form distinguishing a different species.[43]

Based on arbitrary morphological characteristics, Cohn's classification proved inadequate to account for physiological irregularities observed within his species. Having already postponed the task of determining the historical relationships of the "form-genera" and "form-kinds" he and others had proposed, he was nevertheless convinced that bacteria, like all plants and animals, consisted of distinct kinds, and that it was only their extraordinary smallness that concealed this.[44] Cohn's classification system worked well for the larger bacteria, which always appeared in the same forms in endless numbers and without intermediary shapes even in the most varied conditions. His taxonomy was fundamentally challenged, however, by the appearance of constant physiological differences among "morphologically and even essentially (*wesentlich)*" identical forms. Whether these physiological phenomena appeared in the media in which the microbes lived, in the products they

[41] Cohn, "Untersuchungen über Bacterien," 128.

[42] Ibid., 130.

[43] Ibid., 129–130.

[44] Ibid., 133.

produced, or in the characteristics of their movement, they disrupted the integrity of Cohn's taxonomy.[45]

Cohn rejected Pasteur's solution to this problem. Pasteur insisted that the nature of organized ferments (bacteria) could be determined conclusively by studying, not their microscopic structure, but only their physiological functions. Should each microbial form that appeared constantly in a particular medium or that had a characteristic fermenting effect be declared its own species or kind, even if it had not been investigated microscopically?[46] For Cohn, though, Pasteur's path would lead to "purely physiological kinds" which would not be defined morphologically "like 'good' species," but exclusively physiologically."[47] In the future, Cohn argued, perfected microscopy would render physiological criteria unnecessary by revealing new morphological differences that were concealed in apparently identical organisms. This new technology would provide investigators the tools to peer into bacterial structure and thus discover new "primary varieties and kinds."[48] Cohn's devotion to this morphologically-oriented classification scheme reflected his conviction in the stability of microbial species—and Pasteurian microbial physiology shook those convictions.

Struggling to balance his commitment to morphological taxonomy with a flood of new data provided by recent physiological studies, Cohn eventually conceded a restricted role for physiological characteristics in his classification system. After 1872, research on microorganisms had increased to such an extent, that by the time he published his second "Investigation on Bacteria" only 3 years later, a comprehensive review of the literature had become unfeasible. These remarkable research efforts, however, did little to resolve Cohn's fundamental questions, to which he returned in his 1875 report.[49] Although he once again emphasized morphological characteristics in his taxonomy, he devoted more attention to the "biological" questions of growth and nutrition. This led Cohn to create an expanded place for physiology. He still resolutely believed that within the "familial associations" of bacteria (those that appeared morphologically identical) numerous genera and kinds were yet to be discovered.[50] He now distinguished, however, between a microbe's innate (*angeborenen*) characteristics and newly discovered traits (described by recent

[45] Ibid., 134.

[46] Ibid.

[47] Ibid., 134–135.

[48] Ibid., 135. When Cohn discussed bacterial nutrition, a topic that could hardly be avoided after Pasteur's work, he made no explicit references to taxonomic issues. Instead he treated it as a separate issue, just as he had other topics, such as the ability of bacteria to resist heat. Ibid., 191–192.

[49] Ferdinand Cohn, "Untersuchungen über Bacterien II," *Beiträge zur Biologie der Pflanzen* (Breslau: J.U. Kern's Verlag, Max Muller, 1875), Vol. III, 141–204.

[50] It is likely that these "familial associations" are related to the "associations" studied by ecologists such as Warming and Tansley. See Eugenius Warming, "*Plantamsfund: grundtrak of den økologiska plantegeographi* (Copenhagen: Philipsens Forlag, 1895); idem, *Lehrbuch der Ökologischen Planzengeographie. Eine Einführung in die Kenntniss der Pflanyenvereine* (Berlin: Borntraeger, 1896), trans. E. Knoblauch.

physiological research) that appeared in a microbe's response to changing nutritional and other vital conditions.[51]

For Cohn, only these innate traits justified the establishment of distinct species. Because his taxonomy was constructed with such "constantly transmitted characteristics [between generations]," he now considered it to be a "natural" system.[52] Distinguishing between these two kinds of traits allowed him to maintain the integrity of his taxonomic arrangement, while conceding the role of some physiological criteria in defining "independent species (*Species*)". A particular bacterial form could now be considered an "independent" species or kind if it was "constantly bound by characteristic physiological phenomena."[53] Nevertheless, no matter how constantly these physiological characteristics might be expressed by a certain microorganism, Cohn still preferred to classify bacterial species (*Arten*) by means of "external features" observable under the microscope.[54] Cohn's new allowance for physiological characteristics coincided with his efforts to encourage a broader biological perspective in studying microorganisms. Although he considered his taxonomy a "natural" system, it would ultimately fail to account for the biological phenomena that he himself was beginning to discuss and investigate.

Cohn brought this biological perspective to his investigation of peculiar organisms that expressed strange, yet constant, physiological properties—organisms that contained various elemental granules within their bodies and lived in poisonous environments. His approach to bacterial systematics emerges clearly in his treatment of the best known of these organisms, the *Beggiatoa*. (These investigations of "Strongly Refracting Granules in *Bacteria* and *Beggiatoa*" and "The Development of Hydrogen Sulfide by *Beggiatoa*" would prove to be especially influential for Vinogradskii's a decade later.[55]) *Beggiatoa* had attracted wide attention because they, like many red colored and colorless (chlorophyll-less) microbes, contained dark granules in their cells. The nature of these granules remained a mystery—were they integral parts of the cell or simply by-products of living in their environments? Most investigators considered these granules to be the defining characteristic of the species that possessed them.

For Cohn, these granules met the strict criteria of a genuine physiological trait. It had been assumed that *Beggiatoa*'s lack of chlorophyll meant that they could not "assimilate purely inorganic nutrients" (like light), which explained their frequent occurrence in organically rich environments such as the putrefying waters of "muddy, stinking ditches and factory waters."[56] Cohn challenged this

---

[51] Cohn, "Untersuchungen über Bacterien II," 142.

[52] Ibid.

[53] Ibid. Here he mentioned the specific ferment-effects of microorganisms described by Pasteur.

[54] Ibid.

[55] These are section titles in Cohn's "Untersuchungen über Bacterien II"; "Stark Lichtbrechende Körnchen in Bacterien und Beggiatoen" is section 13, 172–173; and "Schwefelwasserstoffentwicklung durch Beggiatoen" is in section 14, 173–174.

[56] Ibid., 172–173.

assumption, however, recognizing that *Beggiatoa* also populated mineral rich sources (thermal springs), which contained only traces of organic matter. It had long been known, Cohn reminded his readers, that *Beggiatoa* were the characteristic inhabitants of sulphur hot springs.[57] At these sites one could not escape the *Beggiatoa*, which grew in huge colonies of "white, mucilaginous masses covering the bottom of the waters, or swimming around in slimy blobs."[58] Cohn's attempt to interpret the role of *Beggiatoa* biologically, and especially his use of physiological criteria, resonated with Vinogradskii as he began to organize his own investigation.[59]

Cohn synthesized new research by the chemist Lothar Meyer and the botanist, and a founder of plant ecology, Eugenius Warming with his own investigations, proposing a new understanding of *Beggiatoa*'s relationship to its environment. These ideas would inform Vinogradskii's experiences—his style of data collection and subsequent interpretations—during his sulphur spring expeditions, and thus deserve a brief review. In 1862, while observing the *Beggiatoa* masses at the bottom of Georges sulphur spring in Landeck, Silesia, Cohn surmised that these organisms generated the hydrogen sulfide gases emanating there. He found support for his notion in Lothar Meyer's 1863 analysis of this Landeck water.[60] Assigned to investigate the thermal springs located around Landeck in the *Graftschaft* of Glatz by the Well Authority, Meyer measured four thermal springs—Georgenquelle, Marienquelle, Wiesenquelle, and Marianenquelle—for their temperature, alkalinity, specific weight, and the production of gases. Following Bunsen's method, Meyer concluded that all four springs did indeed produce hydrogen sulfide gases.[61] In his predominately chemical report he mentioned a living organism in only one place—when discussing *Beggiatoa*'s role in the production of hydrogen sulfide gases. He conducted a simple experiment to explore this phenomenon. After storing Landeck water containing *Beggiatoa* in closed bottles for 4 months, Meyer measured five times more hydrogen sulfide than was normally found in fresh thermal

[57] Cohn's research included trips to thermal hot springs throughout the Pyrenees, Alps, and the Euganeen river basin in Italy.

[58] Ibid., 173.

[59] Winogradsky, "Ueber schwefelbacterian," 492–493.

[60] Today's readers will be familiar with Lothar Meyer (1833–1895) for his formulation of a periodic law of the elements independent of D. I. Mendeleev's work. Meyer trained originally as a physician at the University of Zurich and at Würzburg, receiving his MD in 1853. Encouraged by Justus Liebig, his physiology professor at Zurich he turned to the study of physiological chemistry and studied with Robert Bunsen at Heidelberg. It is interesting that his earliest research concerned the physiological study of the uptake of gases by the blood (1856), work similar to Hoppe-Seyler's (who would later discover hemoglobin in1864). On Meyer, see Otto Theodor Benfey, "Julius Lothar Meyer" *Dictionary of Scientific Biography*, Vol. 9, 347–353.

[61] Lothar Meyer, "Chemische Untersuchungen der Thermen zu Landeck in der Grafschaft Glatz" *Journal fur Praktische Chemie*, Band 91, Heft 1, 1864, 1–14, esp. 5–6. Meyer had recently been assigned the directorship of the chemistry laboratory in the physiological institute in Breslau, not far from Cohn's own laboratory.

water. In Meyer's work Cohn found a corroboration of his own conclusion that *Beggiatoa* produced the hydrogen sulfide content of the springs. He then expanded his investigation of *Beggiatoa*, combining his own laboratory experiments with Warming's observations.[62]

Vinogradskii appreciated Cohn's ability to correlate laboratory observations with those made in natural settings. Cohn had observed *Beggiatoa*'s activities in a variety of settings—in sea aquariums and experimental "Ehrlenmeyer" flasks, and in natural settings such as the sulphur springs—and he also drew on the published observations of other researchers who had become interested in *Beggiatoa*. He was especially struck by Warming's observations on a *Beggiatoa* bloom off the Danish coast that had grown so large that it annoyed the inhabitants along the shores between Copenhagen and Helsingör with its powerful odor of rotting eggs and unnerving silver-black color.[63] The content of Warming's samples matched that of Cohn's sea aquariums—the predominate organisms were *Beggiatoa* and the granules contained in their cells proved to be reguline (pure metallic) sulphur all of which "refracted light strongly."[64]

Based on this collection of observations, Cohn began to ask the questions that would so interest Vinogradskii in 1885. For example, *Beggiatoa*'s ability to develop in such large quantities and still produce hydrogen sulfide, Cohn surmised, required

---

[62] Meyer, following the language of his mentor Robert Bunsen viewed changes in the content of hydrogen sulfide, "proved, as was to be expected, somewhat variable according to the time and circumstances" that samples were taken. Cohn, and Vinogradskii later, may have found the biological significance of this observation enticing. Bunsen when investigation the occurrence of sulphur in coal deposits noted the connection between the great quantity of marsh gas that escaped from the swamp mud was caused by the reduction of the salts of sulfuric acid (sulfates) and the decomposition of humus-like vegetable remains. See Robert Bunsen, "Ueber das Vorkommen von Gyps und Schwefel in Braunkohlenablagerungen," *Studien des Göttingischen Vereins Bergmännischer Freunde*, Bd. IV, 359 (Notizenblatt, Nr. 6) republished in Wilhelm Ostwald, ed., *Gesammlte Handlungen von Robert Bunsen* (Leipsig: Verlag von Wilhelm Engelmann, 1904), Vol. 1, 461–462; See 462. Following Cohn's lead Vinogradskii also read Meyers work. See Sergius Winogradsky, "Ueber Schwefelbacterien," *Botanische Zeitung*, No. 31, 5 August 1887, 489–507, see 491.

[63] Warming (1841–1927) published his own bacterial taxonomy in which he described the *Beggiatoa* Cohn refers to here. See Eugenius Warming, "Om Nogle ved Danmarks Kyster levende Bakterier," (Kjöbenhavn: Bianco Lunos Bogtrykkeri, 1876), Aftryk af *Videnskabelige Meddelelser fra den naturhistoriske Forenin i Kjobenhavn*, 1875, No. 20–28; See esp. 50–59. On Warming see D. Müller, "Johannes Eugenius Bülow Warming" *Dictionary of Scientific Biogrpahy*, Vol. XIV, 181–182. In this work, Warming proposed a new class of *Bacterium sulphuratum* which included all organisms containing sulphur crystals in their cells. He adopted and recommended a taxonic system similar to Ray Lankester's pleiomorphic classification system. In 1895, Warming would also write the first monograph on plant ecology.

[64] Cohn, "Untersuchungen über Bacterien II," 179–180. Vinogradskii would also address the issue of the crystalline nature of the sulphur granules. If the crystals were hard crystalline sulphur they were less likely to be part of the organism's nutritional processes, and were more likely to only an effect of the mechanical, chemical processes occurring around them. This issue was, of course, central to Vinogradskii's physiological approach.

specific environmental conditions. Warming's observations proved that these conditions might exist along the seacoast. Because *Beggiatoa* were not found in such large amounts everywhere where putrefying water existed, Cohn doubted that the same environmental conditions existed in the inland waters.[65] His synthesis of Meyer's, Warming's, and his own research assured Cohn that living organisms were the principle factor in the process of transforming sulfates into hydrogen sulfide.[66] He had found that all of the locations investigated shared two phenomena—a powerful odor of hydrogen sulfide emanated from the waters and dark sulphur crystals appeared in *Beggiatoa's* filaments. His association of *Beggiatoa*'s ability to grow and produce hydrogen sulfide with specific conditions in both laboratory and natural sites offered Vinogradskii a model for pursuing his own objectives in fungal nutrition research.

Using techniques similar to Cohn's, but operating within his own worldview, Vinogradskii offered original solutions to the queries Cohn had left open and challenged those he found erroneous. Although Cohn had found the *Beggiatoa* case to be an intriguing case of physiological constancy, he did not make the intellectual leap to propose a new physiological kind. He risked supposing, however, that the process—the decomposition of sulfuric acid and the production of hydrogen sulfide—that seemed to play a role in *Beggiatoa's* "cycle of vital activities" might be extended to other organisms living in similar environments.[67] Without an experimental investigation of *Beggiatoa*'s nutritional process, he warned, it would be premature to attribute their physiological characteristics to other putrefying organisms[68] Ultimately his commitment to morphological bacterial taxonomy prevented him from identifying a new species characterized solely by its idiosyncratic vital activity.

Trained as a plant physiologist, Vinogradskii did not share Cohn's commitment to morphological characteristics—and here he was not alone. Although Cohn's bacterial taxonomy had become the predominately accepted by the 1880s, it also met resistance in several quarters. The pleiomorphists Lankester, Zopf, and Warming found Cohn's taxonomy too provisional. The bacteria they had investigated exhibited pleiomorphic plasticity too extreme, they argued, to classify them by using Cohn's system of forms and shapes. They suggested alternative taxonomies to replace Cohn's morphological system—taxonomies that recognized purely physiological species or types.

When organizing his Strassburg research, Vinogradskii followed the example of these pleiomorphists, focusing his investigation on bacterial physiology. He was reluctant, however, to completely abandon Cohn's taxonomy because it required species to be constant and yet allowed a role, however restricted, for physiological

[65] For example, there would have to be large amounts of sulfates in the water and an insufficiency of iron (which would combine with the pure sulphur to form other compounds) thus interfering with hydrogen sulfide production.

[66] Cohn, "Untersuchungen über Bacterien II," 1875, 176–177.

[67] Ibid., 176.

[68] Ibid., 177.

processes. At first glance, the more pleiomorphic alternatives to Cohn's system seemed to better fit Vinogradskii's approach by promoting biological, physiological, and chemical understandings of bacteria. His reading of Pasteur under Famintsyn's tutelage, however, led Vinogradskii to view the constancy of species as a fundamental part of the cycle of life. Because each species was assigned to transform (decompose, ferment, or produce) a specific kind of material in nature's economy, they must be stable or there would be no logical coherence in the organization of nature.

Vinogradskii accepted Cohn's classification system because it allowed for investigating bacterial species as biological entities. For Vinogradskii, a biological approach meant correlating the nutritional needs of organisms with their place in nature's cycle of life. According to him, in order to fulfill their particular roles in the cycle of life, microbes necessarily possessed their own consistent life cycles. The novelty of Vinogradskii's approach resided primarily, however, not in his commitment to a particular taxonomic organization, but rather, in his exploration of the physiological identity of a species and the set of techniques he used to discover that identity.

Vinogradskii's research on sulphur bacteria reveals his preoccupation with a physiology founded on a thermodynamic view of life and nature. In St. Petersburg, he had learned to investigate the physiological roles microbes played in the economy of nature by analyzing their nutritional and respiratory processes as dynamic transformations of matter and energy. This commitment is highlighted by his establishment of a sulphur bacteria taxonomy in which he prioritized, not the usual morphological characteristics such as shape and size, but physiological, nutritional processes. He now integrated this approach into De Bary's program of investigating microbial systematics.

De Bary, however, defined the constancy of species in a way that proved awkward for Vinogradskii's approach. This German botanist, who had founded the field of mycology and had contributed much to the study of bacteria, admitted the existence of pleiomorphic species. He maintained, based on observations of changing fungal forms in response to environmental conditions, that a limited plasticity of morphological characteristics occurred in some species.[69] Vinogradskii had a choice to make. If he adhered to De Bary's moderate form of pleiomorphism he would have found it more difficult to classify bacterial species according to Cohn's systematics; that is, by assigning unique identities (names) to single, distinct species based on groupings of forms and processes. By accepting De Bary's view, however, he would have gained some interpretive flexibility. He could discount the slight variation that always appeared bacterial experiments,

[69] The problem of classifying microorganisms relied on demarcating the regular cycles of development that these botanists and mycologists assumed they were observing in nature. This task was especially complicated for the bacteria, which were so small that even under the most powerful of microscopes they appeared as tiny dots. These various views on species constancy clashed most clearly in the taxonomies established by each author.

giving him more flexibility to consider the organisms housed in his laboratory apparatuses as stable, measurable entities.[70] The organisms themselves would decide the issue for Vinogradskii—pushing him from the morphological investigation assigned by De Bary to a physiological one that demanded the skills he had developed in Russia.

[70] S.N. Vinogradskii, *Mikrobiologiia Pochvy: Problemy i Metody, Piat'desiat let issledovanii* A. A. Imshenetskii, ed., (Moskva: Akademiia Nauk, 1951), 74; originally published as Idem., *Microbiologie du Sol: Problèmes et Méthodes, Cinquante ans de Recherches* (Paris: Masson, 1949). To best understand the significance of Vinogradskii's prioritization of physiology one needs to assess the extent to which each of the early bacteriologists relied on morphological versus physiological characteristics in constructing their taxonomies, and to consider how the notion of a 'physiological characteristic' is changing for these scientists.

Chapter 4
Free Nature in the Laboratory

Vinogradskii developed his approach during the laboratory revolution, when scientists increasingly distinguished between lab and field research. Two parallel developments characterized this revolution—on the one hand, laboratory-builders appropriated the natural to authenticate their experimental practices, and on the other, natural historians (field scientists) drew on the authority of heroic adventure to validate their research.[1] Even by the 1890s, when laboratory standards came to dominate the biological sciences, some researchers in nascent disciplines such as microbiology and ecology preserved their natural history perspectives. Vinogradskii, for example, was able to synthesize his laboratory and natural research due in part to his commitment to a conceptual and rhetorical strategy of making "direct observations." For him, viewing an organism's physiological processes microscopically and in artificial cultures gave the data he collected in the laboratory and the field the same epistemological status. For him, organisms developed the physiological characteristics that defined their 'true' roles in the cycle of life in the plenum of nature's complexity. By eliminating that complexity using controlled laboratory environments and careful observational techniques, he believed he could decipher these roles. Vinogradskii's grand vision of the cycle of life—and the methods by which he transformed it into an experimental laboratory investigation—would guide his later research and constitute his great legacy to an international school of soil microbiologists in the twentieth century.

[1] On laboratory-builders see Robert E. Kohler, *Landscapes and Labscapes: Exploring the Lab-Field Border in Biology* (Chicago: The University of Chicago Press, 2002), 3; on heroic adventure see Bruce Hevly, "The Heroic Science of Glacier Motion," in Henrika Kuklick and Robert E. Kohler, eds., *Science in the Field*, *Osiris*, Vol. 11, 1996, 66–86.

L. Ackert, *Sergei Vinogradskii and the Cycle of Life: From the Thermodynamics of Life to Ecological Microbiology, 1850-1950*, Archimedes 34,
DOI 10.1007/978-94-007-5198-9_4, © Springer Science+Business Media Dordrecht 2013

Beggiatoa Nutrition

Soon after beginning the experimental phase of his Strassburg research, Vinogradskii began investigating fungal nutrition and applying the lessons of his apprenticeship. Living in bottles "not entirely hermetically sealed" in a warm room at the laboratory, the *Beggiatoa* had a surprise in store for Vinogradskii.[2] Post-Kochian bacteriology mandated that a successful morphological investigation of bacteria required the use of the pure liquid and gel cultures that had become standard in the 1880s. Vinogradskii followed the prescribed methods suitable for completing De Bary's assignment to describe the life cycles of a few species of microscopic fungi. He initially endeavored to cultivate the *Beggiatoa* samples he had collected on his sulphur spring expedition in several artificial solutions containing easily fermentable materials (*Stoffe*). Surprisingly, he failed to grow any cultures that were healthy enough to proceed with the morphological investigation.[3]

Frustrated by *Beggiatoa*'s inability to grow in the standard cultures, he changed his line of research to investigate their nutrition.[4] He tested how well *Beggiatoa* grew in a series of nutrient solutions and gelatin plates, experimenting with various concentrations of sugar, peptone, ammonium nitrate and other organic materials.[5] He always received poor results. For example, when he cultivated them "in a solution of 1/2% peptone and 1% sugar after 15–20 h the cultures teemed with putrefying bacteria; the *Beggiatoa*, however, disintegrated into small pieces soon disappearing completely."[6] His best result was a culture that remained in a reasonable state for 5 days. Even when he diluted the solutions with well water or water from the nearby Langenbruck mineral spring, the *Beggiatoa* could not compete with the other faster growing bacteria, and "soon fell, dead, to the bottom of the cultures."[7] In general, however, he noticed that the *Beggiatoa* did much worse in the artificial nutrient solutions than in the natural sulphur spring waters.

Vinogradskii did not forget the lessons he had learned during his expedition—using them to explain his experimental observations. In the laboratory cultures, he had concluded, organic material actually hindered *Beggiatoa* growth and simultaneously favored the development of other bacteria, which through their vital activities produced a decomposition product noxious to *Beggiatoa*.[8] This assessment had broader significance for the organism's place in nature's economy: *Beggiatoa's* "beautiful development" in the sulphur springs, he thought, was "easily understandable"

---

[2] On the storage of the *Beggiatoa,* see Winogradsky, "Ueber Schwefelbacterien," 558.

[3] Winogradsky, "Ueber Schwefelbacterien," 569.

[4] Ibid. In his published account, Vinogradskii claimed that he was forced to adopt a physiological approach at this juncture in his investigation. It is clear, however, that his investigation, since its inception, was at its core physiological.

[5] Winogradsky, "Ueber Schwefelbacterien," 556, 569–570.

[6] Ibid., 569.

[7] Ibid., 569–570. The Langenbruck water was used widely in chemical and physiological investigations during this period.

[8] Ibid., 570.

as the result of competition between *Beggiatoa* and other microorganisms. *Beggiatoa* exist in "complete purity" only in the sulphur springs, because there they find the necessary nutrients and "endure no competition (*die Concurrenz*) from other bacteria," which cannot propagate in the poisonous waters.[9] In environments where *Beggiatoa* did live "in society" with putrefying bacteria, the latter, which propagated at a rate much greater than the slow-growing *Beggiatoa*, were able to decompose all the organic nutritive material.[10]

As Vinogradskii's nutrition experiments progressed he relied increasingly on the repertoire of techniques he had developed in Famintsyn's physiological laboratory. Using a complex yet complementary set of cultures—on microscope slides, in retorts, in parallel tests, and in artificial sulphur environments—he hoped to tease apart the nutritional processes of *Beggiatoa* in order to better understand their natural roles in the cycle of life. He correlated his observations of microbial activity in these laboratory cultures with those he had encountered in the literature and during his expedition by applying methods that he vaunted as "direct." Placing organisms from "free nature" into drops of solutions on slides under his microscope ocular, he was, in effect, creating a portal between the natural and artificial, nature and the laboratory, and the particular phenomena of individual organisms and the grand vision of the cycle of life. As he attempted to corroborate and refine his interpretations of *Beggiatoa* physiology these juxtaposed dichotomies shifted in response to his various experimental manipulations.[11] He used his experiments not only to answer specific questions about *Beggiatoa* nutrition, but also to describe a more general natural phenomenon, larger than even the nature of bacterial species.[12]

---

[9] Ibid., 532. He knew form other sources of information that "everywhere else in nature *Beggiatoa* were dependent on the society of other bacteria." Outside the sulphur springs (and in some cases even in them), *Beggiatoa* "must also comply with the conditions of existence" of these bacteria, which provide *Beggiatoa's* necessary nutrients. Does competition in this sense have evolutionary significance?

[10] *Beggiatoa*, could not "endure this competition" in nature except where decomposition produced large amounts gaseous products, such as with the fermentation of cellulose. Ibid., 571. In the summer of 1886, Vinogradskii was partially scooped by Hoppe-Seyler who published his research on the influence of organisms on cellulose decomposition. Hoppe-Seyler discussed *Beggiatoa* nutrition only peripherally, thus Vinogradskii felt that "it was not entirely superfluous if he also reported his investigation in some detail", which was conducted independently and using different methods. Both investigations had come to the same conclusion that (1) *Beggiatoa* filaments play no role in reduction of sulfates or in the forming of hydrogen sulfide and (2) the sulfur deposits in the cells are the result of the oxidation of hydrogen sulfide.

[11] Ibid., 532. Vinogradskii conducted a series of experiments between December 1885 and January of the next year. By 'free nature' Vinogradskii meant the organisms he observed were living outside the influence of artificial conditions, such as he would administer in his "direct" experiments. Here we see the distinction he made drew between nature and the laboratory. The rhetorical use of the notion of "direct" observation or experimentation was widespread during this period. It certainly is related to Pasteur's microbial investigations, but must extend beyond and earlier to that. It is also in Cohn's writing. In this period, it essentially is related to the use of microscopes, which allowed investigators to see into nature in increasingly greater magnifications and detail.

[12] In the next decade, in part due to Vinogradskii's nitrification research at the Imperial Institute of Experimental Medicine in St. Petersburg, Russia, this larger phenomenon would come to be called 'nitrogen cycles' and 'sulphur cycles.'

He was not the first to wonder at the curious relationship between sulphur granules and the filamental bacteria in which they were deposited, and in his 1887 report he presented a historical overview of the work he thought most relevant. Cohn's contribution was first and foremost.[13] In the early 1870s, had Cohn described a group of bacteria that contained sulfur granules in their filaments and were able to live in an environment rich in hydrogen sulfide, a compound poisonous to most other forms of life. Drawing on the work of Adolf Engler and Georg Winter, Cohn agreed that the sulphur granules were the organism's defining morphological characteristic.[14] Later, A. Ehtard and L. Olivier had explained the granules as the organism's chemical response to the presence of sulfates.[15] Zopf thought them merely an indicator of the filament's age—the more granules, the older the filaments.[16] Vinogradskii criticized the conclusion that the granules were simply either surplus reserves or waste material, because he felt that the investigators had not done careful enough experiments, or, he wrote, were based simply on "deductive reasoning and incorrect experimental methods."[17] He did not believe that these sulphur granules were, in fact, simply the morphological characteristics Cohn and others had assumed, not did he accept Zopf's notion that they were merely the result of a mechanical absorption of sulphur from the immediate surroundings.[18]

[13] Cohn's own description (as follows) was adopted in nearly every subsequent retelling, whether recited by those antagonistic or supportive of Cohn's general views.

[14] Vinogradskii cited Adolf Engler, *Ueber die Pilz-Vegetation des weissen oder todten Grundes in der Keiler Bucht* [I have not been able to locate this work]. Georg Winter, "Die Pilze Deutschlands, Oesterreich und der Schweiz," in *Dr. L. Rabenhorst's Kryptogamen-Flora von Deutschland, Oesterreich und der Schweiz*, (Leipzig: Verlag von Eduard Kummer, 1884), Zweite Auflage, Erster Band; On natural systems of fungi see chapter three "System der Pilze" for a discussion of the class *Schizomycetes* (Zopf's *Spaltpilze*), especially 36–38 where he compares Nägeli's and Cohn's genera. He explains that Nägeli described only a very few form-groups (*Formengruppen*), where Cohn established an entire series of genera, which he (Cohn) sharply delineated and broke into a large number of types (*Arten*). Winter claimed a middle ground: on the one hand, there were "some well distinguished and constant genera for example *Micrococcus* (in the wider sense), *Bacillus, Spirillum, and Sarcina*." On the other hand, he also thought that some of Cohn's genera were only "developmental stages", 36. For his description of *Beggiatoa* see 40, 57–59; here he describes them as a genus with nine types.

[15] A. Ehtard and L. Olivier, "De la réduction des sulfates par les êtres vivants," *Comptes Rendus de l'Académie des Sciences*, 1882, 846–849. They criticized those who used the vague denomination "sulfuraires" for these sulphur spring organisms, simply because they "constituent la glarine et la barégine des eaux sulphureauses." In their own investigations they suspected that these "sulphuraires" take in sulphur from the sulfates and give off (*dégagent*) hydrogen sulfide. Ibid., 848.

[16] Wilhelm Zopf, *Die Spaltpilze: Nach dem neuesten Standpunkte bearbeitet* (Breslau: Verlag von Eduard Trewendt, 1885).

[17] Winogradsky, "Ueber Schwefelbacterien," 502.

[18] Ibid., 502–504.

Vinogradskii's Virtuosity: The Role of Sulphur in Beggiatoa Nutrition

By using "direct" observations Vinogradskii believed that his conclusions drawn from laboratory microcultural environments had acquired the same epistemological significance as did those made from observations in "free nature." In order to clarify the 'suppositions' made by previous researchers, he ran a series of experiments using the approach he had developed for his investigation of *Mycoderma vini* during his St. Petersburg apprenticeship. Having failed to learn in which artificial solutions *Beggiatoa* could survive, he began by using the water he had collected along with his samples. Then, using slide microcultures very similar in design to the Geissler chambers of his first investigation, he attempted to elucidate the physiological role of the sulfur granules in *Beggiatoa.* With this apparatus he could observe directly the bacteria's movement as it negotiated the artificial environment he had created. He described his slide microcultures as follows:

> a drop of water containing the microorganism under investigation was placed on a microscope slide. It was covered with a coverslip which is held up at each corner by small pieces of another coverslip in order to provide space for capillary action. It was thus possible to change the liquid in the microculture by placing several drops of liquid on one side of the coverslip using a capillary pipette. Excess liquid was removed by touching the opposite side of the coverslip with strips of absorbent filter paper drawing the liquid across the slide through the space under the coverslip. This action thoroughly washes the microculture.[19]

Using this method he ultimately obtained healthy growths of *Beggiatoa* that he could maintain over prolonged periods of time (up to several months) by washing the microculture several times per day. These regular washings removed the bacteria and infusorians that he knew would outcompete *Beggiatoa* for resources. There was a material reality to consider, too—since *Beggiatoa* filaments attached themselves along their length to the surface of the glass, they were not washed away by the slight flow of the liquid under the coverslip.[20] He conducted a series of tests using this microculture technique, observing the *Beggiatoa* filaments while varying the solution in which they were living. He had observed that when hydrogen sulfide was absent from the microculture the filaments quickly lost their sulfur granules. Having learned how to dissolve the granules it was now possible for Vinogradskii to study the mechanism of their formation.

To determine the origin of the sulphur granules found in the *Beggiatoa* cells he adapted the slide microculture experiment to compare *Beggiatoa*'s reaction to two nutritional conditions simultaneously. The research of Cohn, Meyer, and Zopf told Vinogradskii that the chemical processes occurring in *Beggiatoa*'s natural settings were limited to (1) the reduction of sulfates and (2) the oxidation of hydrogen sulfide.[21] To determine which of these processes caused *Beggiatoa* to undergo physiological

[19] Ibid., 502.

[20] Ibid., 503.

[21] Ibid., 504.

changes, Vinogradskii set up a parallel slide microculture experiment. He placed *Beggiatoa* filaments devoid of sulfur granules into two microenvironments—one rich in sulfates and the other in hydrogen sulfide. To maintain these very sensitive environments he crafted a controllable hydrogen atmosphere and continuously washed the cultures with a sulfate-rich solution.[22]

These techniques gave him the power to control and observe *Beggiatoa* produce and eliminate the granules, the origin of which he quickly discovered. He had observed that in the artificial hydrogen atmosphere sulphur-free filaments were soon filled with countless tiny black dots and after 24 h the same filaments became "crammed-full" with large sulphur granules.[23] Alternately, once removing the *Beggiatoa* from the hydrogen atmosphere their granules soon disappeared. In the sulfate-rich environment, moreover, sulfur granules never appeared in the filaments.[24] He, thus, concluded that *Beggiatoa* produced sulfur granules by oxidizing hydrogen sulfide and not by a mere mechanistic absorption of sulfates. The "extraordinary speed and constancy with which *Beggiatoa* stored sulphur" led Vinogradskii to conclude "that the production of sulphur granules occurred only through the oxidation of hydrogen sulfide."[25] Having tamed *Beggiatoa*'s most intriguing physiological characteristic, he could now expand investigation to address the questions that lay at the heart of his grand vision of nature.

In Vinogradskii's concept of the cycle of life, microorganisms played specific roles as "agents" of chemical transformations. Through processes of decay, fermentation, and oxidation and reduction, they drove the changes that organized nature's building blocks—the elements. These agents determined their niche in the cycle of life by

[22] His deft ability to create artificial environments is demonstrated by his construction of a simple apparatus to maintain the filaments in a hydrogen sulfide atmosphere. Using a large tubular glass bell jar, closed at one end with a cork, through which two glass tubes passed. A wide tube extended deeply to the bottom of the jar with its other end submerged in a small dish containing a calcium sulfide solution, the other smaller tube provided a controllable opening for air to escape when hydrochloric acid was poured into the calcium sulfate solution. He placed his slide culture under the bell jar and proceeded to add the acid—"hydrogen sulfide gas developed inside the sealed bell jar, which gradually diffused into the culture drops." Ibid., 504–505.

[23] The term "crammed-full" (*vollgestopft*) was used by nearly all those who observed this state.

[24] Winogradsky, ", "Ueber Schwefelbacterien," 506. It is interesting that he used Langenbruck sulfur spring water. It was the common choice for experiments probably because Robert Bunsen had provided an analysis of this *Waldquelle* (forest spring). Ibid., 506. See Robert Bunsen, "Anleitung zur Analyse der Aschen und Mineralwasser," *Gesammelte Abhandlungen von Robert Bunsen* (Leipzig: Verlag von Wilhelm Engelmann, 1904), in Wilhem Ostwald and Max Bodenstein *, eds., Auftrage der Deutschen Bunsen-Gesellschaft für angewandte physikalishe Chemie*, Band 3, 500–555. For his gasometric method of analysis see 508–536, and for the table listing the results Vinogradskii used see 555. Vinogradskii used the raw numbers for his calculations, which were averaged in the re-publication of Bunsen's earlier work (see *Zeitschrift für analytische Chemie* (Heidelberg: Carl Winter's Universitätsbuchhandlung, 1874), Op. cit., Bunsen, "Anleitung zur Analyse der Aschen und Mineralwasser," *Gesammelte Abhandlungen von Robert Bunsen*, 1904, 500.

[25] Winogradsky, "Ueber Schwefelbacterien," 505.

their vital physiological processes of nutrition and respiration. This perspective guided Vinogradskii's next experiments on *Beggiatoa* nutrition. Convinced now that *Beggiatoa* required both oxygen and hydrogen sulfide to produce sulfur granules in their filaments, he took what he considered "the next logical step." [26] How did *Beggiatoa* manage their nutritional needs—that is, where did they find enough free oxygen to oxidize hydrogen sulfide?[27] He assumed that *Beggiatoa* must find the oxygen in the water they lived in, however, free oxygen could not exist in hydrogen sulfide-rich liquids. To resolve this contradiction he again resorted to his slide microcultures. Imagining himself in *Beggiatoa*'s environment, he organized a spatial picture describing how *Beggiatoa* negotiated the liquid strata in which they lived. Not limiting his mind's eye to the slides in front of him, he compared observations of three environments—retorts, sulphur springs, and microcultures.[28] In retorts *Beggiatoa* accumulated primarily at the water's surface where they formed tender nets and tufts; however, as they spread out across the retort's glass wall they avoided the surface of the water never growing closer than 1mm.[29] In the sulphur springs *Beggiatoa* behaved similarly—they always congregated just below the surface of the water. On the other hand, neither did they grow in the water's depths—he had never found them at the bottom of any pool deeper than a half meter. In contrast, "they always grew abundantly at the effluence of pools where the water was only a few centimeters deep."[30] Freely correlating his observations in nature and the laboratory, he developed a coherent understanding of *Beggiatoa*'s spatial negotiation of their environment relative to two natural elements—hydrogen sulfide and oxygen.

In the laboratory, the immediacy of his microscopical techniques and apparatuses provided a clear picture of *Beggiatoa*'s preference for strata with less oxygen. His observations of *Beggiatoa* in slide microcultures, he thought, demonstrated how they regulated their own need for hydrogen sulfide and oxygen. When placed into a drop of solution devoid of hydrogen sulfide *Beggiatoa* filaments moved to the center of the drop and formed into a dense ball or tuft.[31] After adding a drop of diluted hydrogen sulfide, the *Beggiatoa* "migrated immediately to the edge of the drop where . . . they formed a thick white border visible [even] to the naked eye."[32]

[26] Ibid., 513–517. Drawing on his own observations and those made by Ehtard and Olivier, and Hoppe-Seyler, Vinogradskii accepted that they required a supply of oxygen.

[27] Vinogradskii had criticized Cohn's neglect of this contradiction. See Winogradsky, "Ueber Schwefelbacterien," 491, 513; Cohn, "Untersuchungen über Bakterien," 1875, 177.

[28] The phenomenon of scientists seeing into nature by imagining themselves as part of it has recently captured the attention of historians. For a good discussion of Barbara McClintock's own "process of integration" see Nathaniel Comfort, *The Tangled Field: Barbara McClintock's Search for the Patterns of Genetic Control* (Cambridge, MA: Harvard University Press, 2001), 67–68.

[29] Winogradsky, "Ueber Schwefelbacterien," 514.

[30] Ibid.

[31] Ibid., 515.

[32] Ibid. Vinogradskii uses a descriptive language very adroitly here, as in other sections of his paper, effectively drawing his readers not only into his argument, but also through his microscope into the secret world of the sulphur bacteria.

This border, Vinogradskii observed, always grew 1 mm from the edge of the drop. If there were enough filaments in the drop they would even create a solid wall between the slide and coverslip. Upon microscopic investigation, these *Beggiatoa* borders struck him as appearing "very peculiar":

> the filaments form a thickly woven mass stretching parallel to the coverslip. They are pressed thickly together and creep to and fro, in and out. Other [filaments] travel across (*durchkreuzen*) the mass in various directions, twining themselves around one another and the straight filaments in the most whimsical manner. Rising … from this netting filamental tails and curves oscillate to and fro, at times retreating into the mass of filaments and later emerging again—it is an extremely elegant and animated image.[33]

For Vinogradskii the elegance of this image carried significance well beyond the *Beggiatoa* border. Watching the bacteria form a ring-like border in an intermediary position between the hydrogen sulfide water and the free oxygen of the atmosphere, he interpreted their oscillating movements as an exchange of gases between these organisms and their environment. Here he was looking at this phenomenon through the lens provided, not by De Bary's morphological approach, but rather, by Famintsyn's physiological thermodynamics. At this stage in his investigation Vinogradskii imagined a correlation between the activity of the sulfur bacteria filaments (their intricate negotiations of strata) with their nutritional and respiratory processes—as they moved inside the border they absorbed hydrogen sulfide and when outside they oxidized it. Thinking physiologically, he imagined that the *Beggiatoa* were "regulating this two step process according to their own [nutritional] needs."[34] For him this was no less a respiratory act than was an animal's inhalation of oxygen and exhalation of carbon dioxide, or, a plant's absorption of carbon dioxide and emission of oxygen.

Again relying on 'direct' experimentation, he endeavored to corroborate this was indeed bacterial respiration. To prove that *Beggiatoa* obtained their vital energy from hydrogen sulfide and oxygen, he constructed a series of parallel slide microcultures to test how well *Beggiatoa* would grow in various environments. In his most significant experiment he set up three slides with access to free oxygen, washing them three times daily—one with hydrogen sulfide and two with plain water. After 2 months the *Beggiatoa* appeared in "an excellent state" only in one slide, the one with a hydrogen sulfide environment—"its filaments were moving lively and were filled with sulphur granules."[35] Eventually, in fact, the *Beggiatoa* grew to such an extent that the slide became so "crammed-full" with filaments that he had to end the experiment.[36] Vinogradskii took *Beggiatoa*'s exuberance as proof that they needed oxygen and hydrogen sulfide to *live*. He now made the connection that when they were most healthy *Beggiatoa* were also filled with sulphur. This revelation brought him back to a question that had intrigued him even prior to his Alpine excursions,

[33] Ibid.

[34] Ibid., 515–516.

[35] Ibid., 532.

[36] Ibid., 532–534.

having emerged from his reading of Cohn's 1875 report. What, precisely, was the significance of the sulphur granules for *Beggiatoa*?

The nature of *Beggiatoa*'s sulphur—a riddle that had preoccupied Vinogradskii for the past 2 years of experiments, during countless hours of peering through his microscope as he washed slide after slide—could now be investigated as a nutritional process. Now knowing that *Beggiatoa* needed hydrogen sulfide to live, Vinogradskii addressed the "kernel" of the entire question: Why did *Beggiatoa* need so much sulphur and what was its nutritional significance?[37] Drawing on his ability make *Beggiatoa* filaments accumulate sulphur granules, he measured the maximum amount of sulphur they could endure. He determined that not only could the filaments contain up to 95% of its specific weight in sulphur, but that they were able to eliminate and replenish this amount every 24 h. [38] This incredibly rapid cycling of sulphur, however, did not correlate with a rapid increase in *Beggiatoa* growth—at best they doubled their length in 24 h. These facts negated his initial supposition that *Beggiatoa* use the sulphur to build their bodily structures, and so led him to another conclusion. Through a series of microchemical tests he determined that *Beggiatoa* not only oxidize hydrogen sulfide (thus forming the sulphur granules), but continue the oxidation process to the next stage. Once the sulphur is deposited in their filamental cells they then oxidize it into sulfuric acid (thus eliminating the granules).[39]

Vinogradskii's investigation of *Beggiatoa* nutrition—initiated to answer a simple morphological question—thus, led him eventually to a provocative conclusion with much broader implications: *Beggiatoa* obtained their vital energy through a complex physiological process. They first assimilated sulphur into their cells by oxidizing the hydrogen sulfide found in their environment—from sulphur spring waters or the putrefaction of organic matter—with atmospheric oxygen. This assimilation process produced sulphur reserves in their filaments, which they then oxidized into sulfuric acid. Vinogradskii compared this nutritional process to respiration in higher plants, a process by which organisms gained the matter and energy needed to support their vital processes.

A New Physiological Type

Vinogradskii, then, fulfilled the second part of De Bary's initial assignment, to consider a new taxonomy for microscopic fungi—but in a quite different manner than De Bary had in mind. Applying the model of his *Beggiatoa* research to other organisms that he considered similar physiologically, he proposed an entirely new,

[37] Ibid., 546–547.

[38] Ibid., 547.

[39] Ibid., 548.

physiological type of organism.[40] His research on these other organisms convinced him that all filamental bacteria containing sulphur granules possessed the same physiological processes. Drawing a new picture of that corner of the microbial world, he established a taxonomic group based on a physiological type. Naming this group the "sulphur bacteria," he reclassified many of the organisms described by previous researchers (Zopf especially) as their own bacteria species or simply as stages in outlandish pleiomorphic life cycles. Ironically, Vinogradskii organized his new taxonomy around the same characteristic all other systems had used—the sulphur granules. For him, however, they were not a morphological trait, but a sign that an organism was fulfilling a particular, and constant physiological process in the cycle of life.[41]

The intensive labors of his microscopic fungi research in Strassburg produced other scientific products in addition to his "Ueber Schwefelbacterien" (1887). During this research, he encountered another group of bacteria earlier investigators had classified in the same genera as *Beggiatoa*.[42] Morphologically the organism (*Leptothrix*) seemed closely related to *Beggiatoa*—both were filamental and accumulated dark crystals in their cells. Vinogradskii found, however, that *Leptothrix* demonstrated quite different physiological processes. Most significantly, the dark crystals turned out to be ferrous oxide (iron) and not sulphur. Continuing his sulphur work, he initiated an investigation into *Leptothrix* nutrition. In his opinion *Leptothrix*'s physiological characteristics varied enough from *Beggiatoa*'s to constitute their own physiological type—and he termed them "iron bacteria."[43] As with his sulphur bacteria research, the monomorphism-pleiomorphism debates provided a backdrop for this investigation. He contrasted his results, for example, to Zopf's who considered *Leptothrix* to be a more advanced stage than *Beggiatoa*, and the ultimate stage in the complex life cycle of the *Spaltpilze*.[44] Running parallel to his *Beggiatoa* research, the iron bacteria project confirms Vinogradskii's commitment to a vision of the cycle of life, in which nature could be studied as a thermodynamical system of physiological processes regulated by microbial nutrition and respiration.

---

[40] Ibid., 608. The long list includes the purple bacteria Lankester, Zopf and Warming had classified as stages of a pleiomorphic species called *Bacterium rubescens* (Lankester), *B. sulfatum* (Warming) and *Beggiatoa roseo-perscina* (Zopf). See S.N. Vinogradskii, "Dannye o morfologii i fiziologii serobakterii" in *Mikrobiologiia Pochvy: Problemy i Metody, Piat'desiat let issledovanii*, 84–120; see 84–89; originally published as Sergius Winogradsky, *Zur morphologie und Physiologie der Schwefelbacterien* (Leipzig: Arthur Felix, 1888).

[41] Vinogradskii followed previous classification trends by subdividing his new physiological type (sulphur bacteria) according to morphological characters, such as filament thickness.

[42] Ibid., 588–589. He first believed that the filamental bacteria *Cladothrix* (called *Leptothrix* by Zopf) contained sulphur granules in its filaments, but he had, he later found, deceived himself—they were iron oxide granules that had lodged in the bacteria' sheath.

[43] Sergius Winogradsky, "Ueber Eisenbacterien," *Botanische Zeitung*, No. 17, 27 April 1888, 261–270, see 261.

[44] Zopf, *Die Spaltpilze*, 4, 11.

Vinogradskii had developed a unique system for investigating microorganisms physiologically, one that he applied to *Beggiatoa*, *Leptothrix* and subsequently to many others. Following the same logical sequence he had with *Beggiatoa*, he developed experiments to demonstrate that *Leptothrix* oxidized ferrous oxide to satisfy its nutritional needs. He also established a classification system for his second physiological type—filamental bacteria whose life-processes were characterized by the ability to oxidize ferric protoxide salts—under the general term iron-bacteria.[45] He was not content, however, merely to describe the morphological association between *Leptothrix* and iron. As always, his commitment to a vision of the cycle of life guided his interests and research, driving him to seek the iron's physiological significance to *Leptothrix*.

From his sulfur bacteria and iron bacteria research Vinogradskii generalized about the physiological peculiarities of all microorganisms. For him these two microbial groups were similar because they both possessed the ability to absorb a mineral, oxidize it, and then release the byproduct into the surrounding environment. Extending this comparison to other microorganisms, he would construct a new category of vital activity—chemosynthesis.

The Reception of Vinogradskii's Research

Vinogradskii's research at Strassburg attracted immediate and significant attention in the rapidly expanding microbiological community. Several reviews of his work appeared, written primarily by scientists with a stake in the monomorphism-pleiomorphism debate or in the organisms Vinogradskii discussed. The botanist Moritz Büsgen, who was antagonistic towards Zopf's pleiomorphic classification system, applauded Vinogradskii's results and especially his methods of research. Büsgen stressed Vinogradskii's use of day-long and week-long continuous observations of single specimens in slide cultures, which he considered "the only proper basis for establishing [a bacterial] developmental history."[46] Theodor Engelmann, well-known at the time for his research on the photosynthetic capabilities of microorganisms, drew on Vinogradskii's sulphur bacteria work to support his own critique of recent purple-bacteria classification schemes.[47] Focusing his attack on the usual

[45] Protoxides are any of a series of oxides that contains the lowest proportion of oxygen. Iron bacteria shared certain morphological traits: they are all covered by an ochre colored sheath, which gains that color due to being evenly impregnated with ferric hydroxide.

[46] Moritz Büsgen, a review of "S. Winogradsky, *Beiträge zur Morphologie und Physiologie der Bacterien*, Heft I. *Zur Morphologie und Physiologie der Schwefelbacterien* (Leipzig: Arthur Felix, 1888)" in *Botanische Zeitung*, Vol. 47, nr. 1, 4 Jan. 1889, 14–16; 15 for quote.

[47] Theodor Wihelm Engelmann, "Die Purpurbacterien und ihre Beziehungen zum Licht," *Botanische Zeitung*, Vol 46, No. 42, 19 October 1888, 661–669; Idem, No. 43, 677–687; Idem., No. 44, 693–701; Idem., No. 45, 709–720; Vinogradskii is referenced on 663, 695–696, 715, 719. Vinogradskii had concluded that *Beggiatoa* needed only a very little amount of organic matter to live, which supported Engelmann's view that they could live by photosynthesis.

suspects—Lankester, Warming, and Zopf—Engelmann found collaboration for his argument that purple bacteria obtained their vital energy from photosynthesis. Thus codified in the literature, interest in Vinogradskii's work in this area would be available in the early-twentieth century when these bacterial groups were found to play important roles in soils and agricultural research.[48]

Vinogradskii's sulphur bacteria investigations were the first stage of a research program that would define a crucial aspect of his scientific legacy.[49] They would lead him into the next phase of his career to discover chemosynthesis, a new physiological process by which organisms were able to live on inorganic materials. During his Strassburg research Vinogradskii practiced the skills he had leaned with Famintsyn in St. Petersburg. Specifically, he synthesized the cycle of life with a laboratory based experimental program organized around a strong inventory of flexible techniques and a conceptual and rhetorical strategy of direct observation. Vinogradskii had developed the theoretical approach and basic experimental design (the direct method) that he used in his Strassburg work with Famintsyn. During his German internship, however, he was able to apply this Russian training to new questions under the mentorship of one of Europe's foremost botanists. The mature research style he developed in these two schools, would lead him to path breaking discoveries of chemosynthesis, the nitrogen cycle, and a new ecological method in soil microbiology.

In 1888, Vinogradskii left De Bary's laboratory to search for an academic position in the Russian university system. Failing to find a position either in St. Petersburg or Kiev, in the fall he returned to Strassburg.[50] When he arrived he learned that De Bary was suffering from a serious cancer in his jaw. Even though De Bary's face was deformed by an operation to remove the cancer, Vinogradskii recalled that "[f] or some time [De Bary] still came to the laboratory, tried to keep his spirits up, glanced into the microscope, but there was no doubt that he was a finished man. De Bary was soon taken to the grave."[51] De Bary's death led Vinogradskii to a new environment that would also have an impact on his scientific development and legacy. He decided to leave Strassburg, visiting his previous homes in Kiev and St. Petersburg before settling finally in Zurich, Switzerland in the fall of 1888.

---

[48] For one example, see the type of agricultural microbiology being developed at the Rothamsted Agricultural Experiment Station by Sir John Russell, Daniel Cutler, and H. Thornton; See Chap. 5.

[49] In my analysis of Vinogradskii's research during the second half of the 1880s, I will focus on the aspects of his research and life that would bring him to the attention of the world, and would become the center of his future investigations, from the discovery of chemosynthesis to ecological soil microbiology in the 1920s. One of these sides of his work was his development of the so-called "direct experiments," that is, a collection of methods used primarily in the chemical and botanical investigations at the microscopic level. This approach would reappear in his investigations in varied forms throughout his career and would dramatically impact ecologically-minded pedologists in the 1920s–1940s; See Chap. 4.

[50] V.L. Omelianskii, "Sergei Nikolaevich Vinogradskii: Po Povodu 70-letniia so dnia Rozhedeniia," *Arkhiv Biologicheskix Nauk*, Vol. XXVII, Nos. 1–3, 10–36; see 18.

[51] Ibid., 17.

Part III
Ecology

Chapter 5
Vinogradskii's Transformation from Plant Physiologist to Ecologist, 1890–1920

Between the 1880s and the 1930s Vinogradskii underwent a striking intellectual transformation. In 1888, he considered himself a plant physiologist concerned with investigating the nutritional requirements of microorganisms. He conducted his research within a theoretical framework he called the cycle of life. By the 1930s, he identified himself as a soil microbiologist; and although he continued to study microbial nutrition, he now investigated it as an ecological phenomenon.

As he moved to new institutions and took up new research questions throughout this transformation, he maintained his commitment to studying the role of microbes in the cycle of life. During this period, he worked at the Swiss Polytechnic Institute in Zurich where he studied nitrogen bacteria, developed a new, "elective culture" method, and discovered autotrophism—the ability of microbes to live solely on inorganic matter (1888–1890). In 1891, Vinogradskii moved to the Imperial Institute of Experimental Medicine in St. Petersburg, where he expanded his nitrogen work into one of the first programs of soil microbiology (1891–1912). From 1905 to 1919, he moved steadily into retirement and took up scientific farming on his Kiev estate. After the abrupt curtailment of his retirement, he moved to Belgrade University where he taught courses in agricultural microbiology and founded a Department of Soil Science (1920–1921). It was his at next and final position—at the newly founded laboratory of Agricultural Microbiology at the Pasteur Institute in France—that he completed his transformation into an ecologist.

We need to explore his development as a scientist against the background of the rise of ecology, the rapid expansion of Russian soil science, and the laboratory revolution. Historians have portrayed the history of ecology as a late nineteenth century reorientation of Humboldtian plant geography in the context of Darwinian evolutionary theory. In this story, German, British, and American botanists from the 1860s onward studied plant relationships in terms of the struggle for existence and natural selection. My story in this chapter, however, adds a new dimension to this portrayal. Vinogradskii developed into an ecologist by synthesizing, not Humboldtian plant geography and Darwinian evolution, but microbiology and the concept of the cycle of life.

L. Ackert, *Sergei Vinogradskii and the Cycle of Life: From the Thermodynamics of Life to Ecological Microbiology, 1850-1950*, Archimedes 34,
DOI 10.1007/978-94-007-5198-9_5, © Springer Science+Business Media Dordrecht 2013

Autotrophism

In 1889–1891, Vinogradskii conducted the research that would launch his career. The work he had begun in Strassburg on nitrogen bacteria and their role in the cycle of life led him to discover autotrophism—"a new kind of life," as he called it, in which microbes subsist on only inorganic materials. Autotrophism provided a new mechanism for understanding how matter migrated between nature's inorganic and organic realms in terms of the activity of living beings.[1] Vinogradskii's research altered fundamentally the way many scientists understood the history of the earth, focusing their attention on the role the smallest living beings played in forming the environment. This new understanding of microbial nutrition contributed a significant new perspective to a number of scientific disciplines that formed or expanded at the end of the nineteenth century, including microbiology, soil science, and geobotany (Russian plant ecology).

Vinogradskii's move from De Bary's botanical and mycological laboratory to the chemistry laboratories of Zurich initiated his personal transformation from plant physiologist to microbiologist and soil scientist. Transferring from a botanical laboratory to an institutional context dominated by chemistry led Vinogradskii to shift his program of research. In Zurich, he switched the objects of his investigations from filamental sulphur and iron bacteria to non-filamental nitrogen bacteria. The first organisms live solely in water; the latter exist in many environments, especially, and most significantly for Vinogradskii, in the soil.[2] A second phase of this transformation occurred when he moved to St. Petersburg in 1891 and expanded his nitrogen bacteria research. In both these contexts, the demands of studying nitrogen bacteria introduced changes in his repertoire of laboratory techniques and subsequently improved his ability to integrate his laboratory experiences into his global perspective. He continually strove to synthesize his investigations—through the language of methodology, systematics, and physiology—within the cycle of life concept, a task for which his nitrogen bacteria research proved exceptionally fortuitous.

In 1889, Vinogradskii extended his investigation of new physiological types to the question of nitrification. Moving from De Bary's mycological laboratory to the chemistry laboratories in Zurich, he brought with him the cycle of life perspective that he had refined and deepened in his research on sulphur and iron bacteria.

---

[1] In 1897, the plant physiologist Wilhelm Pfeffer incorporated Vinogradskii's nitrification research in his discussion of plant nutrition. He discussed Vinogradskii's work in his section on "The Production of organic Substances through the Assimilation of Carbonic Acid," which he subdivided into: A. Photosynthetic assimilation and, B. Chemosynthetic assimilation. Pfeffer's word chemosynthesis would eventually come to designate Vinogradskii's concept of bacteria autotrophism, even in the latter's own writing. See Wilhelm Pfeffer, *Pflanzenphysiologie: Ein Handbuch der Lehre vom Stoffwechsel und Kraftwechsel in der Pflanze* (Leipzig: Verlag von Wilhelm Engelmann, 1897), 2nd Edition, Vol. 1, 346–349; 385, 547–548.

[2] Taking up the study of nitrogen bacteria also brought Vinogradskii to soil science, the field in which he would make his most lasting scientific contributions.

Applying this approach to new experimental objects—non-filamental nitrogen bacteria—he asked new questions and advanced new experimental methods. Vinogradskii adjusted his investigation to address the nutritional requirements of his new organisms: while sulphur and iron bacteria flourished in very specific environments, nitrogen bacteria populated many environments. To exploit the glimpse of the new world order he had discovered in Strassburg, he hoped to apply new sophisticated laboratory methods to capture novel observations on the nature of nitrogen bacteria. Intuition, he said, told him that this organism would provide at least another example of physiological type, and possibly much more.[3] Based on his earlier studies of sulphur and iron bacteria, in which he used his discovery of their peculiar nutritional demands to classify several new microbial species, he proposed a novel concept of 'the physiological type.' He employed this conceptual framework in his nitrogen bacteria work in Zurich, essentially redefining the debate over the nature of nitrification. Moreover, this work provided another example of peculiar microbial physiology, which, in combination with his previous work, led him to propose a novel and astounding physiological law of nature. To accomplish this he refined his skills in chemistry and developed an array of new laboratory techniques.

Vinogradskii moved to Zurich with a plan. In Strassburg, it had become evident to him that to pursue an investigation of inorganic physiology he needed to improve his chemistry skills.[4] He was well aware of the city's reputation as a vibrant hub for scientific activity and as a grand center for chemistry. There he worked in three laboratories at the Swiss Polytechnic Institute: in Arthur Hantzsch's analytical chemistry laboratory, Ernst Schultz's physical chemistry laboratory, and at Otto Roth's bacteriological laboratory.[5] The amount of time Vinogradskii spent in each of these laboratories reflects his changing research priorities at a time when he increasingly styled himself a microbiologist in the French tradition—preferring Pasteur's approach of correlating chemical processes with certain microbial agents in determined environmental conditions. Accordingly, he devoted himself to studying organic chemistry, flirted with physical chemistry, and only begrudgingly worked in the only bacteriological laboratory available in Zurich—Roth's. Vinogradskii disliked Roth's approach to bacteriology, which was based on Robert Koch's pure-culture methods.[6]

[3] Serge Winogradsky, "Recherches sur les organismes de la nitrification," *Annales de l'Institut Pasteur*, 1890, Vol. IV, No. 5, 220–221.

[4] Georgii A. Zavarzin, "Sergei Nikolaevich Vinogradskii, 1856–1952," E. N. Mishustin, ed., *Khemosintez: K 100-letiiu otkrytiia S. N. Vinogradskim* (Moskva: Nauka, 1989), 5–21; See 13–15 on Vinogradskii in Zurich. This article was translated into English and published as, Idem, "Sergei N. Winogradsky and the Discovery of Chemosynthesis," Hans G. Schlegel and Botho Bowien, eds., *Autotrophic Bacteria* (Madison: Science Tech Publishers, 1989), 17–32.

[5] Although he later denied that the directors of these laboratories influenced his work in any significant way, it seems very likely that Vinogradskii's increased sophistication in chemistry can be attributed only to this Zurich experience. His disavowal can be ascribed to priority claims for discovering chemical autotrophism. Much of his scientific legacy—especially for founding an ecological microbiology—was based on his 1890s papers.

[6] Vinogradskii probably chose these laboratories because their directors were studying nitrogen compounds.

Vinogradskii had mixed emotions about his time in Zurich. On the one hand, the city's chemistry institutes impressed him because it offered no less than three professors of chemistry, provided ample space for students and visitors, and had "every instrument one might need."[7] On the other hand, he found working in large and heavily populated laboratories a daunting experience, which made him feel "very small and unimportant."[8] Under Hantzsch's wing, Vinogradskii increased his proficiency at chemistry sufficiently "to proceed to more complex chemical manipulations."[9]

By 1890, Hantzsch was on his way to becoming one of Europe's top chemists. His studies of the stereochemistry of organic nitrogen compounds and the integration of physico-chemical methods of measurement in analytical chemistry contributed to a new understanding of the relationship between inorganic and organic compounds.[10] By the end of the century, this research would lead chemists to doubt the validity of the border that most of them had believed to separate inorganic and organic compounds.[11] The trend in chemistry represented by Hantzsch's research may well have appealed to Vinogradskii who, while in Hantzsch's laboratory, was attempting to break down the same boundary in biology.

Vinogradskii complemented his lessons in Hantzsch's laboratory by studying physical chemistry with Ernst Schultze.[12] Not only did Vinogradskii find Schultze to be "a modest and amiable man who gave [him] a free hand to do whatever he wanted in the laboratory," he shared Schultze's interest in nitrogen research. Schultze's nitrogen work was a contribution to physiological and agricultural chemistry, which was primarily concerned with understanding how and in which phases organic substances are formed from carbonic acid, ammonia, and water.[13] Through his research

---

[7] Zavarzin, "Sergei Nikolaevich Vinogradskii, 1856–1952," 13. Here Zavarzin drew on unidentified autobiographical notes written by Vinogradskii around 1952, these were not available to me, for some unknown reason, when I conducted research at the Archives of the Russian Academy of Sciences, Moscow. Recently, and posthumously, Zavarzin published extensive extracts of this autobiography in G.A. Zavazin, *Tri zhizni velikogo mikrobiologa: dokumental'naia povest' o Sergee Nikolaeviche Vinogradskom* (Moskva: Librokom, 2010).

[8] Ibid. Vinogradskii maintained this preference for small laboratories for the duration of his career.

[9] Ibid.

[10] On his life and work see Joachim Stocklöv, *Arthur Rudolf Hantzsch im Briefwechsel mit Wilhelm Ostwald,* Vol. 21, *Berliner Beiträge zur Geschihte der Naturwissenschaften und der Technik* (Berlin: Ellen R. Swinne-Verl., 1998), 18–52. One of the works that led to his wide influence on the history of chemistry was Arthur Hantzsch, *Bemerkungen über sterochemisch isomere Stickstoffverbindungen* (Berlin: A. W. Schade's Buchdr., 1890).

[11] Stocklöv, *Arthur Rudolf Hantzsch im Briefwechsel mit Wilhelm Ostwald,* 37.

[12] On Vinogradskii in Ernst Schulze's laboratory see Zavarzin, "Sergei Nikolaevich Vinogradskii, 1856–1952," 13–14. Schulze applied physical chemistry methods to investigate agricultural problems. He published several articles on a variety of subjects: Ernst Schulze, "Sur les matériaux de réserve et plus particulièrement sur les tannin des feuilles persistantes," *Annales agronomiques*, Vol. 14, 1888, 525; idem, "Les engrais verts en sols pauvres," *Annales agronomiques*, Vol. 21, 1895, 394–396.

[13] Ernst von Meyer, *A History of Chemistry from the Earliest Times to the Present Day, Being also an Introduction to the Study of the Science* (London: Macmillan and Co., Ltd., 1898), Second Edition, trans. George McGowan, 536.

on the production of nitrogenous compounds during such processes as the germination of seeds he provided some of the foundational work in phyto-chemistry.[14] Under Schulze's guidance, Vinogradskii learned how to determine all forms of nitrogen chemically, "in order to study effectively the problem of nitrification."[15] In this work, he developed the skills in analytical chemistry and instrument building that would allow him to advance his research agenda.

To investigate nitrification, which Vinogradskii considered a biological phenomenon, he needed more than chemistry skills. Thus, as congenial, well-organized, and well-supplied as he found the chemical laboratories to be, he had to forego these comforts. There was only one place in Zurich, he recalled, where he could pursue his goals—Otto Roth's bacteriological laboratory.[16] Vinogradskii would come to hate this laboratory, which he considered "as ugly as its boss."[17] At the root of his animosity lay his dislike for Roth's strict adherence to the standard bacteriological procedures developed by Robert Koch. Roth had abandoned a short career as a physician to study with Koch in Berlin.[18] Immediately upon graduation in 1890, Roth became the first professor of hygiene and bacteriology at both the Institute of Hygiene at Zurich University and the Polytechnic Institute. Fresh from Koch's school, he taught short courses in bacteriology for physicians at the Zurich University based on Kochian principles. His rigid adherence to a standardized syllabus—and one must assume a reluctance to accept any criticism of those methods—earned Vinogradskii's ire.[19] Yet, based on the experiments he conducted in this laboratory, Vinogradskii created the novel "elective culture" experimental method for investigating microbial nutrition, used that method to discover autotrophism, and launched a career in microbiology and soil science.[20]

Nitrification as a Biological Phenomenon

Vinogradskii's nitrification work in Zurich marks the beginning of his transition from plant physiology to soil microbiology. Nitrification—the study of how ammonia is transformed (oxidized) into nitrates and nitrites in nature—became a pressing

[14] Ibid., 537–538.

[15] Zavarzin, "Sergei Nikolaevich Vinogradskii, 1856–1952," 13.

[16] Ibid.

[17] Ibid., Op cit.

[18] There is little available information on Otto Roth.

[19] This anger was fresh in Vinogradskii's mind when he recounted his Zurich days; even after a long career, he had "never seen a professor to be such an ass either before or after." Zavarzin, "Sergei Nikolaevich Vinogradskii, 1856–1952," 13; Again Zavarzin is drawing on an unspecified autobiographical document.

[20] In the 1920s, Vinogradskii would work to create an independent discipline of soil microbiology. He also hoped to found soil microbiology on the approach that he would develop from the elective culture method of his Zurich research.

topic of scientific research in the late nineteenth century, especially after it had been shown to be a biological phenomenon. When Vinogradskii took up the question, however, scientists had yet to reach any consensus about the nature of this phenomenon.[21] Agricultural chemists, plant physiologists, and microbiologists—with their different disciplinary criteria and methodologies—still debated whether to investigate it as a chemical, biological, or physical process. Vinogradskii, with the experience of his sulphur and iron bacteria research, his new chemical training in Zurich, and his background in Russian plant physiology engaged the question from a novel perspective.

Trained as a plant physiologist, and with developed microbiological interests, Vinogradskii initially investigated nitrification as a biological phenomenon. Although he was not the first to do this, the previous biological investigations of nitrification failed to satisfy his criterion of identifying physiologically the precise organism responsible for nitrification.[22] Applying the approach that he learned from Famintsyn, Vinogradskii set out to explain the mystery of nitrification according to the vital activity (nutrition and respiration) of specific microbiological agents.

When writing up his first nitrification investigation, Vinogradskii portrayed his work as a logical extension of his Strassburg research. His previous discovery of two new physiological types of microbes that were able to subsist on inorganic matter—an idea clearly informed by his commitment to the cycle of life—motivated him to seek out the "elusive agent of nitrification."[23] He presumed, he recalled, that "if there are organisms like sulphur bacteria, which can oxidize hydrogen sulfide, then one with good reason may suggest the existence of specific organisms that live on such a rich source of energy as the oxidation of ammonia."[24] It is possible, however, that he also chose nitrification as a subject because of its practical applications to sanitation and agriculture.

---

[21] The source of Vinogradskii's initial interest in nitrification is unclear. Possibly, as Zavarzin thought, it was likely that Vinogradskii's interest was aroused by the polemics surrounding Guyon and Dupetit's denitrification work. These French agricultural chemists succeeded in isolating the organisms responsible for denitrification—a process that many feared might lead to soil infertility during the application of fertilizers. Vinogradskii might have thought that a better understanding of the reverse process, nitrification, would prove helpful. See Zavarzin, "Sergei Nikolaevich Vinogradskii, 1856–1952," 12.

[22] Serge Winogradsky, "Recherches sur les Organismes de la Nitrification," *Annales de l'Institut Pasteur*, Vol. 4, 1890, 215–231; 257–275, 760–811, idem., Vol. 5, 1891, 92–100, 577–616. He also abstracted these articles for presentations to the French Academy of Sciences—See Serge Winogradsky, "Sur les Organismes de la Nitrification," *Comptes Rendus des Seances de l'Académie des Sciences*, Vol. 60, No. 19, 12 May 1890, 1013–1016, and Idem., Vol. 63, No. 2, 1891, 89–92. See also Alfred Koch's review of "Serge Winogradsky, Recherches sur les Organismes de la Nitrification," *Botanische Zeitung*, Vol. 49, Nr. 40, 2 October 1891, 669–672, 680–685, 698–701.

[23] Serge Winogradsky, "Recherches sur les Organismes de la Nitrification," 215.

[24] Ibid., 215–216.

In the 1870s, Jean Jacques Schloesing and A. Müntz first investigated nitrification as a biological phenomenon.[25] As Professor of Agricultural Chemistry at the newly founded Institute National Agronomique in 1875, Schloesing was assigned to determine the feasibility of transporting sewage from Paris to the French countryside for irrigation purposes.[26] With Müntz, he set out to investigate whether the presence of humus (organic matter consisting primarily of decaying plants) in the soil was necessary to purify the sewage water. Defining purification as complete nitrification, they measured how long it took for the ammonia to disappear. Surprised by the long delay before oxidation even began (20 days) they surmised that nitrification was not controlled by access to "active oxygen" (a chemical explanation) but rather by the action of intermediary "organized ferments."[27] To test this, they used boiling temperatures and chloroform gas to prove that nitrification would not occur in conditions deadly to living beings.[28] The final statement of their first nitrification report stated simply that it remained "to discover and isolate the nitrifying organisms." Their best efforts over the next year using Pasteur's methods failed to describe these organisms any more specifically than by the vague term "*ferment nitrique*."[29]

Between 1879 and 1881, Robert Warington, an agricultural chemist at the Rothamsted Agricultural Experiment Station in England, approached nitrification as an agricultural problem. He not only confirmed the conclusions of Schloesing and Müntz, but also went a step further. He established that soil nitrification was a two-stage process in which bacteria first convert ammonia into nitrites and subsequently into nitrates.[30] Although Warington had studied advanced bacteriological techniques with the Austrian bacteriologist Eduard Emanuel at London's Brown Institution, he failed to isolate the nitrifying organisms.[31]

---

[25] In 1877, Jean Jacques Theophile Schloesing and A. Müntz demonstrated that nitrification is a biological process by using chloroform vapors to inhibit the production of nitrates. See J. Schloesing and A. Müntz, "Sur la Nitrification par les Ferments Organisés," *Comptes Rendus de l'Académie des Sciences*, 1877, Vol. 84, No. 7, 301–303; Idem., Vol. 85, No. 22, 1018–1020; Idem., 1878, Vol. 86, No. 14, 892–895; Idem., 1879, Vol. 89, No. 21, 891–894 ; Idem., No. 25, 1074–1077. On the contributions of Schloesing and Müntz to the history of microbiology see: Raymond N. Doetsch, *Microbiology: Historical Contribution from 1776 to 1908 by Spallanzani, Schwann, Pasteur, Cohn, Tyndall, Koch, Lister, Schloesing, Burrill, Ehrlich, Winogradsky, Warington, Beijerinck, Smith, Orla-Jensen* (New Brunswick: Rutgers University Press, 1960), 103–107; and Selman A. Waksman, *Principles of Soil Microbiology* (Baltimore: The Williams and Wilkins Company, 1932), Second Edition, 62–63. For a discussion of the relationship between soil chemistry and plant growth see Aaron J. Ihde, *The Development of Modern Chemistry* (New York: Harper & Row, 1964), 420–426.

[26] Doetsch, *Microbiology: Historical Contributions from 1776 to 1908*, 103.

[27] J. Schloesing and A. Müntz, "Sur la Nitrification par les Ferments Organisés," Vol. 84, No. 7, 302.

[28] Ibid., 303.

[29] Idem., Vol. 89, No. 21, 892.

[30] Waksman, *Principles of Soil Microbiology*, 63.

[31] Doetsch, *Microbiology: Historical Contribution from 1776 to 1908*, 156–157.

A. Hereaus took up this line of research and in 1886 claimed that he had succeeded in isolating the organisms.[32] In soil, water, and air cultures rich in ammonia and organic nitrogen, he succeeded in detecting nitrification—the production of nitrates and nitrites through the oxidization of ammonia or nitrogen. Using standard pure culture methods, he isolated the bacterial species living in his cultures, two of which, he found, produced intensive nitrification. Vinogradskii questioned the sterility of Hereaus's cultures and pointed out that the nitrification Heraeus had detected was likely due to "an inadvertent contamination of sulfates and sulfites, which float in the air at all times."[33] Vinogradskii's survey of this literature taught him that nitrification was a biological problem and that the study of soil microbes had practical value. As the decade progressed, he would develop these insights into a new area of research—soil microbiology.

Investigating nitrification led Vinogradskii to develop a novel, "elective culture" method in microbiology. The specificity and flexibility of this method would make it the centerpiece of his technical repertoire, and would re-emerge in the 1920s as an ecological method. He developed this method because of his frustrating failures to induce a strong nitrification process in the laboratory. Conducting a long series of tests using standard bacteriological methods, Vinogradskii—like the above investigators—also failed to isolate any nitrifying organisms.[34] His experiences with growing successful cultures of sulphur and iron bacteria inspired him to avoid the standard methods and devise a new approach based on what he considered "a strictly inductive method."[35] First, he would search for the liquid culture most favorable to nitrification. Second, he would carefully maintain these cultural conditions while conducting a series of transplants designed to eliminate the species not adapted to those conditions. Third, having purified the species content of the culture as much as possible—yet maintaining an intense level of nitrification—he could isolate the nitrifying species. Subsequently, he would be able to test the nitrifying abilities of each isolated species in pure cultures.[36] Following this plan, Vinogradskii, attempted cultivations in several solutions but failed to achieve significant nitrification.

To collect his experimental organisms Vinogradskii returned to wild, free nature. Avoiding the cultures already available in the Zurich laboratories, he instead collected samples of two different Zurich soils. Using his new skills in analytical and physical chemistry he inoculated a variety of solutions with these little clumps of complex nature, attempting to create a "liquid nutritive environment," which produced a strong nitrification. His efforts—adjusting for alkalinity and for different kinds of ammonia

[32] A. Hereaus, "Sur les Bactéries des Eaux de Source et sur les Propriétés Oxydantes," *Zeitschrift für Hygiene*, Vol. 1, 1886, 193–234.

[33] Winogradsky, "Recherches sur les Organismes de la Nitrification," 215–216.

[34] Ibid., 220–221.

[35] Ibid. I mention this simply because it is quite clear that it was in fact a strongly deductive method. He had a very good idea of his objective—nitrifying microbes. He adjusted his media recipes to home in on these organisms.

[36] Winogradsky, "Recherches sur les Organismes de la Nitrification," 222.

salts—produced only a very faint nitrification.[37] "Then the idea came to [him]," he wrote, "to eliminate all organic matter, [and this] immediately resulted in an intensive nitrification in his cultures."[38] Because of his sulfur and iron research, this fact did not greatly surprise him. By relating the action of nitrification to his previous work, he had made a breakthrough—not only was organic material unnecessary for nitrification, it actually impeded it. Henceforth, he grew his cultures on a simple mixture of "very pure water" from Lake Zurich and *inorganic* ammonia salts.[39]

Up until this point in his nitrification work, Vinogradskii had been working primarily as an agricultural chemist. The key to solving the enigma of nitrification was the keen observational ability he had developed during his training as a plant physiologist and microbiologist. In his investigation, he combined macroscopic and microscopic observations to explore the nitrifying environment he had created in the laboratory. He had observed that the layer of ammonia salt on the bottom of his liquid culture was finely divided into two layers: a darker one and a perfectly white one.[40] He noticed that as his cultures aged, this pure layer "changed strangely"; it became grayish in color and gelatinous in consistency.[41] Microscopic observation of this peculiar layer revealed transparent lumps of the salt covered with "thick groups of beautiful oval bacteria."[42] He strongly suspected that he had found his agent of nitrification.

Vinogradskii developed the elective culture method during the final "isolation" step of his new approach. His task was to eliminate the four additional species that were living in his culture, and thus, that might play a role in nitrification. Tediously purifying the surfaces of his retorts—distilling the water twice and adding sulfuric acid to it, calcinating all of his inorganic salts and his cotton stoppers—he succeeded in producing an absolutely pure mineral culture medium.[43] Conducting these manipulations in conjunction with careful microscopic observations of bacterial behavior, Vinogradskii isolated his "beautiful ovals" and named them *Nitrosomonas*.[44] Isolating this organism proved to him the power and efficiency of his new method.

---

[37] Ibid.

[38] Ibid.

[39] Ibid., 223.

[40] Ibid., 226.

[41] Ibid.

[42] Ibid.

[43] Winogradsky, "Recherches sur les Organismes de la Nitrification," 269.

[44] The beautiful ovals he first noticed at times became somewhat elongated. Spindle-shaped cells with blunt ends appeared in the population and sometimes came to be the predominate form. Its cells were for the most part motionless, but periodically went into motion. This period of motion occurred suddenly and its intensity caused the culture liquid to grow turbid with swarms of cells. When at rest, the cells sat in thick conglomerations in the form of amorphous zoogloea. The microbes affixed themselves to grains or flakes of carbonates, which dissolve over time as a result of the microbes' action. The bacteria's growth is dependent on direct contact with the carbonic salts of alkaline-soil metals.

Vinogradskii's commitment to the cycle of life predisposed him to extrapolate from his laboratory experiments to challenge fundamental assumptions of general physiology. Although he accepted that his predecessors Schloesing and Müntz had found a special organism related to nitrification, he did not initially accept their term "*ferment nitrifique*."[45] The idea implicit in that term—that a single organism was responsible for "exercising that function over the entire surface of the globe—seemed to him improbable. In its place, he proposed using a broader classification—one based on "physiological type"—because under it one could classify numerous species or varieties.[46] He suggested that this new physiological type, representing an entire cast of nitrifying organisms, existed not only in Zurich's soils but also in all soils on the Earth.[47]

This global perspective was no mere rhetorical flourish. Vinogradskii's vision of the cycle of life now generated a comparative study of nitrifying organisms around the world. Drawing on his international contacts, he requested soils from Europe and other "exotic sources" such as Africa, Asia, South America, and Australia.[48] He tested the organisms in the soil samples for their relative ability to oxidize ammonia into nitrites in both liquid cultures and directly in their original soils. His found that these organisms—from diverse localities—oxidized ammonia at the same rate. Nitrification, he concluded, was a universal, biological phenomenon that transcended local conditions and thus existed everywhere.[49]

The nitrification research provided Vinogradskii with his third example of an inorganically based respiration from which he divined what he called a new physiological fact—the fact of autotrophism. His experiments had shown "that living beings could accomplish a complete synthesis of organic matter on our planet independent of the solar rays."[50] That is, since they contained no chlorophyll, these

[45] He did eventually use this term after he isolated the two species of microbes that participated symbiotically in nitrification—calling them *ferments nitreux* and *ferments nitriques*. See Serge Winogradsky, "Recherches sur les Organismes de la Nitrification," Vol. 5, 1891.

[46] Serge Winogradsky, "Recherches sur les Organismes de la Nitrification," Vol. 4, 1890, 230; Idem., 1891, 593.

[47] In order to refute the doctrine "held by physiologists world round," that organisms can survive without organic nutrients, one would have to provide very weighty proof. For 3 months, he successfully cultivated them in a solution, which was excruciatingly prepared to be devoid of organic matter (including boiling all his glassware twice in sulfuric acid). To prove definitively the nature of nutrition he measured the amount of organic carbon in the sediment of his cultures. If carbon appeared during the experiment, only *Nitrosomonas* could have produced it. He varied the cultures by adding different forms of nitrogen to them. In all cases, carbon appeared in the test cultures, proving, he thought, that *Nitrosomonas* possessed the ability to assimilate carbon from carbon dioxide.

[48] He requested samples from Emil Duclaux, Cramer (Zurich), Treub (Buitenzorg), and Cavalcanti (Campinas). From Europe he used soils from Zurich, Gennevilliers, Kazan, and his home town of Podolia (He had investigated these two Russian soils previously, in the winter of 1889–1890.) He received his African soils from La Reghaia, Rouiba, Mitidja, and Tunis. The Asian soils were sent from Buitenzorg, Java, and Tokyo, Japan. The American soils were from Campinas, Brazil, and Quito, Ecuador. The Australian soil came from Melbourne. Idem., 1891, 581.

[49] Ibid., 593.

[50] Serge Winogradsky, "Recherches sur les Organismes de la Nitrification," Vol. 4, No. 5, 1890, 269 and 275.

organisms could not obtain energy from light to drive their oxidation processes.[51] Autotrophism, then, was the process by which organisms developed in a purely mineral environment, acquiring their only source of nutrition from some single inorganic material. In these investigations, he found new experimental evidence to support his cycle of life perspective.

In the above investigations, he accumulated information and experience which led him to develop his idea of autotrophism. For each organism that he studied, he generalized to a broader classification; from *Beggiatoa* to sulfur bacteria, from *Leptothrix* to iron bacteria, and from *Nitrosomonas* to azotobacteria. During the nitrogen studies, he became convinced that the analogy he recognized between these three physiological types could be extended to all of microbiology. From his investigation of sulphur, iron, and nitrogen bacteria, Vinogradskii now confidently extended his vision of the cycle of life to a law of nature. A substantial number of species did not require what most physiologists considered the necessary ingredients for life. Not only did new organic matter arise from the life processes of living beings through photosynthesis, he argued, but also as a result of autotrophism. This notion swept through the scientific community and was adopted by scientists in a broad spectrum of disciplines including plant physiology, soil science, and eventually ecology. It also had the immediate effect of launching Vinogradskii on a career in microbiology.[52]

Soil Microbiology at the Imperial Institute of Experimental Medicine

The widespread acceptance of autotrophism earned Vinogradskii a number of job offers from prestigious institutions. The president of the Zurich Polytechnic Institute invited him to deliver a course of lectures as a privatdozent. In 1891, the Russian immunologist Elie Metchnikov brought Vinogradskii an invitation from Louis Pasteur to set up a bacteriological laboratory at the newly founded Pasteur Institute. Vinogradskii received a comparable offer from the newly established Imperial Institute of Experimental Medicine (IEM) in St. Petersburg.[53] After long deliberation, Vinogradskii accepted the Russian offer and in 1891 moved there to head the laboratory of general microbiology.

Vinogradskii's first trip to St. Petersburg to meet with the founder of the IEM, Prince Oldenburgsky, set the exasperating tone that characterized much of his

[51] Ibid., 275. Although he did not apply the term chemosynthesis to this phenomenon until later, the concept—that a general physiological condition existed, in which organisms did not need light or oxygen to live—was implicit in his description.

[52] Sergei Vinogradskii, *Mikrobiologiia Pochvy*, 169.

[53] For a recent discussion of the conditions at the Imperial Institute of Experimental Medicine during Vinogradskii's tenure, see Todes, *Pavlov's Physiology Factory*, 20–40.

15-year career there. Writing to Metchnikov in July 1891, just after accepting the Prince's offer, Vinogradskii described his tour of the grounds and his negotiations for laboratory space.[54] A small wooden building housed laboratories that were in embarrassingly poor condition. Conversations with the other scientists already there and with the Prince himself did little to clarify plans for new construction. Vinogradskii took the initiative and formulated his own, impressing the Prince enough to place an architect at Vinogradskii's disposal.[55]

Vinogradskii's few demands reveal his scientific style. He requested a "tolerable laboratory, where three or four could work comfortably: him, an assistant, and two praktikanty (outside investigators)."[56] Even more important to him than procuring this modest amount of space were issues of control. He spent considerable effort to ensure that he would be able to work in peace (*spokoino*) and have unrestricted power to manage his own laboratory.[57] Vinogradskii tested his authority 6 months later when he became upset that construction on his new laboratory had not yet begun. Reflecting his personal style, he went straight to the source and brought his demands directly to Prince Oldenburgsky—2 weeks later he was ready to begin research.[58]

In addition to editing the IEM's journal, training a small number of students, and satisfying the bureaucratic and social duties associated with working for his patron, a prince in the Russian royal family, he continued to develop a program in soil microbiology. Utilizing the IEM's extensive resources, Vinogradskii converted his global conceptualization of physiological types and autotrophism into a laboratory investigation of the cycle of life. In Zurich, using novel physiological methods, he had confirmed that nitrification was indeed a two-stage process. Moreover, he had isolated the microbes responsible for each stage. He would now apply his approach to other soil processes closely related to nitrogen bacteria.

Worried about Vinogradskii's professional future, Metchnikov had encouraged him to take up the study of pathology in 1890. By 1893, however, Vinogradskii was committed to soil microbiology. His new direction of research—nitrogen fixation—had "opened a very wide field" and he was getting a flood of ideas.[59] He was amazed at the apparent limitlessness of the topic—so little had been touched. Furthermore,

[54] Letter from Vinogradskii to Elie Metchnikov of 1 July 1891, in G.E. Reikhberg, ed., *Bor'ba za Nauku v Tsarskoi Rossii* (Moskva and Leningrad: Gosudarstvennoe Sotsial'no-Ehkonomicheskoe Izdatel'stvo, 1931), 156.

[55] Ibid., 156–157.

[56] On the role of praktikanty as the unexpected labor force at the IEM, see Daniel P. Todes, *Pavlov's Physiology Factory* (Baltimore: The Johns Hopkins University Press, 2001), 27–32.

[57] Letter from Vinogradskii to Elie Metchnikov of 1 July 1891, *Bor'ba za Nauku v Tsarskoi Rossii*, 157.

[58] Letter from Vinogradskii to Elie Metchnikov dated 17 November 1891, *Bor'ba za Nauku v Tsarskoi Rossii*, 160.

[59] Letter from Vinogradskii to Elie Metchnikov dated 9 February 1894, *Bor'ba za Nauku v Tsarskoi Rossii*, 167.

that year he hired a very capable assistant, Vasilii Omelianskii, who was a student of Vinogradskii's old chemistry professor Menschutkin. With this optimistic outlook, a well-appointed laboratory, and a devoted collaborator, Vinogradskii pursued a series of topics that bridged general microbiology and soil science, and provided the foundation for his program in soil microbiology.

Vinogradskii's transformation into an ecological thinker coincided with the emergence of ecology as a discipline in Russia. We will see in Part IV that the general approach and laboratory methods that he had elaborated for soil microbiology during the 1890s would become recognized in the 1920s as the beginning of ecological microbiology. The relationship between his development as an ecological thinker and the rise of ecology as a discipline is complex. Through his soil microbiology he contributed concepts (autotrophism) and methods (elective cultures) fundamental to Russian ecology; however, he worked outside of its theoretical framework. His commitment to the cycle of life distanced him from the Darwinian and Humboldtian precepts of ecology's founders. In addition, in the late nineteenth century, microbiology had a limited role in soil science and geobotany, the disciplines from which Russian ecology emerged. Exploring Vinogradskii's relationship to the rise of Russian ecology reveals another dimension of the history of that discipline.

Russian ecology emerged along similar lines as in Western Europe and the United States, where botanical geographers introduced Darwinian evolution into their science. In Russia, however, the development of ecology was distinguished by its close ties with soil science.[60] Historian of Russian geobotany Khans Trass observed that: "[t]he elements of geobotany penetrated into Russian science" in two principal ways. On the one hand, it grew "from practical agriculture and forestry" and on the other; it developed from the unification of two previously distinct sciences: botanical geography (the categorizing of florae according to their geographical location) and the geography of plants (the mapping of geographical regions according to their vegetation)."[61] These two sciences converged to form geobotany, which developed quickly over the 1880s and became an independent science distinguishable from other botanical disciplines by its terminology and methods by the end of the century.[62] Russian geobotany (read, ecology) derived its distinctive characteristics during the 1890s, when it was integrated with Russian soil science.[63]

---

[60] On the relationship of Russian geobotany to American ecology, see Khans Trass, *Geobotanika: Istoriia i Sovremennye Tendentsii Razvitiia* (Leningrad: Izd.-vo Nauka, 1976), 91.

[61] Ibid., 23–24. Trass opposes the views of B. Bykov, that "geobotany appeared as a result of the demands of practical life, based on the botanical-geographical ideas of Russian and Western European scientists." As evidence, Trass notes that the geobotanists Tetsman, Cherniaev, and Filipchenko hardly considered the ideas of botanical geography in their investigations and the geographical botanists Borshchov and Ruprecht paid little attention to the geography of plants.

[62] Among the founders of Russian ecology, Trass includes: S.I. Korzhinskii, A. N. Krasnov, I. K. Pachoskii, G. I. Tanfil'ev, and P. N. Krylov. Their work was continued by another generation of ecologically minded botanists, which included G. F. Morozov, A. IA. Gordiagin, and V. N. Sukachev. Ibid., 28.

[63] Ibid. 24.

In the 1890s, Russian soil science was experiencing a period of great expansion, characterized by the ascendancy of the Dokuchaev school, the formation of new institutes, and the organization of large research expeditions.[64] During this period, soil scientists were staking out their disciplinary boundaries and striving to balance their commitments to natural history with a growing interest in experimentation and laboratory-based investigations. They were attracted to Vinogradskii's concept of autotrophism and nitrification research, which appeared in the early 1890s, seeing them as a way to make their science more experimental and to fulfill the central tasks of understanding the formation of soils.

To understand how Vinogradskii's microbiology contributed to soil science, we need to briefly review its history. From the 1870s–1890s, three founders of Russian soil science—R. V. Rizpolozhenskii, P. A. Kostychev, and V. V. Dokuchaev—disagreed about the conceptual framework and methodological approach of their science, and even about the very nature of the soil.[65] By the end of the century, Dokuchaev—in terms of the number of his students, the popularity of his scientific concepts, and his institutional authority—had come to dominate that science. His interest in autotrophism, Kostychev's research on humus and soil microbes, and Rizpolozhenskii's discussion of geobiological processes created a place for Vinogradskii's research in soil science.

The least well known of these three soil scientists, Rafail Vasil'evich Rizpolozhenskii (1860–1918), criticized Dokuchaev's view of the soil as "an independent natural body and a function of all soil formation factors."[66] In the early 1890s, Rizpolozhenskii suggested, rather, that the soil formed due to the interaction of only two primary factors: "organisms and rock," limiting the other factors to the status of "external conditions."[67] He considered the essence of soil formation to be the "the circulation of elements between organisms and nature, or "the seizure (*zakhvat*) of food by organisms from the unorganized environment and its reciprocal return (*obratnyi vozvrat*)."[68] In this way, organisms would "metamorphose" rock into soil, thus creating an environment, in which they could survive.[69] Based on this concept, he

[64] On the history of soil science I have drawn on: I. V. Ivanov, *Istoriia otechestvennogo pochvovedeniia: razvitie idei, differentsiatsiia, institutsializatsia. Kniga pervaia, 1870–1947* (Moskva: Nauka, 2003); I. A. Krupenikov, *Istoria pochvovedeniia: ot vremeni ego zarozhdeniia do nashikh dnei* (Moskva: Nauka, 1981); This work was translated into English: as I. A. Krupenikov, *History of Soil Science: From its Inception to the Present* (Brookfield, VT.:A. A. Balkema Publishers, 1993), trans. A. K. Dhote; and Dan Yaalon and S. Berkowitz, eds., *History of Soil Science: International Perspectives* (Reiskirchen: Catena Verl., 1997).

[65] Here I ignore Nikolai Mikhailovich Sibirtsev (1860–1900), who also contributed in essential ways to the founding of soil science in Russia. His work did not, it seems, extend the role of microbiology beyond what Dokuchaev had imagined. For a short review of his work, see Krupenikov, *Istoria pochvovedeniia*, 183–190.

[66] Ivanov, *Istoriia otechestvennogo pochvovedeniia*, 102.

[67] Ibid.

[68] Krupenikov, *Istoriia pochvovedeniia*, 186.

[69] Ivanov, *Istoriia otechestvennogo pochvovedeniia*, 102.

attempted to establish a new scientific discipline that he called "geobiology." Informed by Jean-Baptiste Lamarck's definition of the biosphere as the distribution of life on the surface of the Earth (1802), Rizpolozhenskii proposed that geobiologists would study the influence of organisms in the "zones of interaction between living and nonliving, and organic and inorganic environments."[70] Although soil scientists rejected much of his work, many—including Dokuchaev's student, V. I. Vernadsky—reacted favorably to his view of "soil formation as the interaction of organisms and rock," what he termed "geobiological processes."[71]

In the early 1870s, Pavel Andreevich Kostychev (1845–1895) promoted making the study of the soil an independent discipline separate from geology and geography. Along with other agronomists, he thought it necessary to study the soil as "an environment for vegetation and for their concrete properties (such as, humus, structure, nutritive materials, etc.)"[72] He considered questions of soil genesis, and the compilation of soil maps to be useless activities for agronomists and more appropriate for geologists.

It was Kostychev who first introduced microbiology into soil science. From 1870 to 1893, he wielded considerable authority in soil science—teaching courses in agriculture, soil science, and plant growing at the Forestry Institute; and at times at the St. Petersburg University. Between 1883 and 1891, he held administrative positions of agriculture inspector (1883–1891) and then director (1893–1895) in the Ministry of Agriculture and State Property.[73] In the middle of his career, the government dispatched him and the founder of Russian microbiology, L. S. Tsenkovskii, to Europe to study at the laboratories of Pasteur and Koch. Sent there to study the production of anthrax vaccines, Kostychev took an interest in Pasteur's work on soil microbes. Upon his return to Russia, Kostychev initiated some of the first investigations of soil microbiology in Russia.[74] His research provided the first experimental proof that microscopic fungi contributed to the formation of humus in the soil.[75] He asserted that the formation of "chernozem (black earth) is a botanical issue."[76] He characterized soils by the accumulation and cycling of humus, or decaying organic matter. In his schema, bacteria and fungi played the central role in this

[70] Ibid.

[71] Ibid., 102–103.

[72] Ibid., 94.

[73] Ibid., 97; and Krupenikov, *Istoriia pochvovedeniia*, 175.

[74] Ibid., 97, 99. For an example of Kostychev's work in this area, see his "Sostav organicheskikh veshchestv poch (peregnoia) v sviazi s voprosom o poleznosti mikorits" in P.A. Kostychev, *Izbrannye trudy* (Moskva : Izd-vo Akademii nauk SSSR, 1951), ed. I.V. Tiurina, 363–366.

[75] Ibid.; N. M. Sibirtsev, "Pamiati P. A. Kostychev: K godovomu dniu konchiny Pavla Andreevicha Kostycheva," in P. A. Kostychev, *Pochvy chernozemnoi oblasti Rossii: Ikh proiskhozhdenie, sostav i svoistva* (Moskva: Gosudarstvennoe Izd-vo Sel'skokhoziaistvennoi Literatury, 1949), 193–197; see 195; and I. S. Kossovich, "Kratkii ocherk rabot i vzgliadov P. A. Kostychev v oblasti pochvovedeniia i zemledeliia," in Idem., 198–223; see 201–202.

[76] Krupenikov, *Istoiria pochvovedeniia*, 176.

decomposition.[77] By creating a place for microbiology in soil science during its formative period, Kostychev prepared the way for Vinogradskii's soil microbiology in the 1890s.

During the 1870s–1890s, Russian soil scientists witnessed a debate between two of the founders of their science over its fundamental concepts. From his earliest days, Kostychev "irreconcilably and inventively, and quite often not objectively" challenged Dokuchaev on nearly every issue—theoretical or organizational—related to soil science.[78] Kostychev's animosity towards Dokuchaev arose in 1876, when the Free Economic Society (VEO) selected Dokuchaev instead of him to organize expeditions to the chernozem (black earth) regions. Only in the last 2 years of Kostychev's life did the pair overcome the personal, social, and scientific differences that divided them.[79]

Of the three figures, Dokuchaev (1846–1903) contributed most to the founding of scientific soil science.[80] He was the first to conceptualize the soil as "an independent natural-historical body" and a system, to create methods for studying it, and to develop the first system of laws for understanding the organization and formation of soil "zones."[81] He outlined the theoretical and practical significance of soil science for society and trained two generations of soil scientists and geobotanists. Through his teaching, his publications, his expeditions, and the voluminous data they produced he set the foundation upon which soil science developed, not only in Russia, but around the world.

During his trips to investigate Russia's chernozem (black earth) regions in the 1870s, Dokuchaev had developed a natural historical conception of the soil.[82] In 1879, he had defined the soil as "the surface mineral-organic formation … that originates due to the interaction of living and dead organisms, the matrix [or gangue] of rock, the climate, and the topography (*rel'ef*) of the terrain."[83] By 1886, he expanded his definition to include the idea that the soil forms from rock "naturally changed by the reciprocal action (*vzaimnoe vliianie*) of water, air, and various types of organisms, living and dead, which have a definite composition, structure, and color."[84] Prior to Vinogradskii's discovery of autotrophic microorganisms, Dokuchaev relegated microorganisms and animals to a role secondary to that of plants, which

---

[77] Ibid., 177.

[78] Ivanov, *Istoriia otechestvennogo pochvovedeniia*, 95; and Idem., *Pavel Andreevich Kostychev* (Moskva, Nauka, 1987).

[79] Ivanov, *Istoriia otechestvennogo pochvovedeniia*, 95. The Agricultural and Economic Department of the VEO (Vol'noe Ekonomicheskoe Obshchestvo) was located in the Department of Agriculture of the Ministry of State Property (Ministerstvo Gosudarstvennyx Imushchestv). Ibid., 103–104.

[80] For a comprehensive list of work on Dokuchaev's life and scientific activity see Krupenikov, *Istoria pochvovedeniia*, 151–153.

[81] Ivanov, *Istoriia otechestvennogo pochvovedeniia*, 46–49.

[82] Ibid., 48.

[83] G. F. Kir'ianov, *Vasilii Vasil'evich Dokuchaev, 1846–1903* (Moskva: Izd-vo Nauka, 1966), 39; and Ivanov, *Istoriia otechestvennogo pochvovedeniia*, 59–61.

[84] Krupenikov, *Istoriia pochvovedeniia*, 161–162.

were able to "synthesize organic matter."[85] In the 1890s, Vinogradskii's research convinced Dokuchaev that autotrophic microorganisms play a primary synthesizing role in soil formation.[86]

Dokuchaev envisioned a role for microbiology in his reform of soil science training. In 1891, Dokuchaev participated in a commission that studied higher education in agriculture. He found the only active organization—the Novo-Aleksandriiskii Institute of Agriculture and Forestry—to be "in a condition of complete disorganization."[87] The following year he became director of the institute and introduced educational reforms that included teaching physics, chemistry, geology, soil science, and plant physiology.[88] These courses produced Russia's first, well-trained specialists in the areas of agriculture and forestry.[89] In an attempt to extend his reforms to the university system, Dokuchaev recommended that the government establish chairs in soil science and microbiology.[90]

For Dokuchaev, Vinogradskii's research represented a significant advance in applying bacteriology to agriculture and soil science. Following the great, though overly theoretical work of Pasteur and Cohn, Vinogradskii (and a series of other bacteriologists) had developed new methods to investigate geological processes in terms of the vital activity of bacteria.[91] Especially important, Dokuchaev thought, was the role of bacteria in the nitrogen cycle, that is, in Vinogradskii's ongoing research at the IEM. Referring to Vinogradskii's work on nitrification, Dokuchaev felt confident that "without bacteria, plants would not be able to use the scanty nitrogen reserves in the soil."[92] His familiarity with Vinogradskii's work is also evident in Dokuchaev's program for a university course in microbiology.[93] At the top of his list of the most important "physiological bacterial types," Dokuchaev placed: the nitrifying organisms, sulphur bacteria, iron bacteria, and nitrogen fixing bacteria—the past and present targets of Vinogradskii's research.

---

[85] Ivanov, *Istoriia otechestvennogo pochvovedeniia*, 48.

[86] Ibid.

[87] Kir'ianov, *Vasilii Vasil'evich Dokuchaev, 1846–1903*, 63.

[88] Krupenikov, *Istoriia pochvovedeniia*, 165.

[89] Kir'ianov, *Vasilii Vasil'evich Dokuchaev, 1846–1903*, 64.

[90] V. V. Dokuchaev, *K. voprosu ob otkrytii pri Imperatorskikh Russkikh Universitetakh kafedr' Pochvovedeniia i Ucheniia o mikrorganismakh (v chastnosti, Bakteriologii)* (S.-Peterburg: Tipografiia E. Evdokimova, 1895).

[91] V.V. Dokuchaev, "Znachenie uchenie o mikroorganizmakh," *K. voprosu ob otkrytii pri Imperatorskikh Russkikh Universitetakh kafedr' Pochvovedeniia i Ucheniia o mikroorganismakh (v chastnosti, Bakteriologii)*, 49–54.

[92] Ibid., 51. Here Dokuchaev drew on Vinogradskii's description of nitrification's two stages of conversion: from ammonia to nitrite, and nitrite to nitrate.

[93] V.V. Dokuchaev, "Programma (primernaia) universitetskogo kursa ucheniia o mikroorganizmakh," *K. voprosu ob otkrytii pri Imperatorskikh Russkikh Universitetakh kafedr' Pochvovedeniia i Ucheniia o mikroorganismakh (v chastnosti, Bakteriologii)*, 56–61. Although he admitted getting the program for this course only from an unspecified "well-known specialist in the given branch of natural science," it is probable that he received it from Vinogradskii.

Vinogradskii, then, developed his program for soil microbiology during the rise of Russian soil science. His research in Zurich had made clear to him that the study of soil microbes could be applied to practical questions in sanitation and agriculture. He now drew on that experience to contribute to soil science, which had been commissioned by the Tsarist government to come to the aid of Russian agriculture. At this time, microbiology could offer little of practical value to soil science in the field. Vinogradskii's contributions came, thus, in the form of laboratory methods for understanding processes involved in soil formation, such as cellulose decomposition and nitrogen fixation. These methods and the research he conducted with them became increasingly valuable to Russian soil scientists.[94] It was in the context of their incorporation of his methods and ideas that Vinogradskii's work first moved into the realm of ecology.

The growth of Russian geobotany occurred during a period characterized by the rapid growth of capitalism, including in the agricultural sector, and by the popularity of Darwin's theory of evolution.[95] After the 1861 reforms, lands were increasingly developed as commercial zones, especially in southern Russia's black earth areas, leading to agronomic problems and a demand for solutions. During this period, the Russian government organized numerous expeditions to various areas of the country to collect data on soil and botanical resources. Russia's soil scientists and geobotanists organized these expeditions for practical objectives, but the experiences led scientists to develop new views about the interrelationship between soil and vegetation. Coming to view vegetation as one of the many important factors influencing the formation and development of the soil, they would eventually organize new ecological sciences.

By the 1890s, many of Russia's 'men of the sixties' had assumed positions in the state bureaucracy, bringing with them their faith in the power of science. Recognizing the threat of widespread famine, they convinced the Russian government of the great need for understanding and improving the health and fertility of its agricultural regions. The strife during Russia's "Hungry Years, 1891–1892," reached its highest level during the summer of 1891, which experienced "the most serious crop failure since the 1830s" leaving its "great, black earth granary … stand[ing] empty."[96] There were two general causes of the famine. Meteorological conditions—an early onset of winter, a summer drought—had led to bleak harvests.[97] In addition, government

---

[94] During his tenure at the IEM, Vinogradskii also applied his microbiological research skills to medical and health questions. In particular, he investigated the nature of a plague that had appeared in south eastern Russia. I do not treat this story in full because it does not relate directly to Vinogradskii's development as an ecologist. It may help, however, to understand how ecological microbiology relates to epidemiology and medical bacteriology. See Chap. 9 for a discussion of how Rene Dubos applied his training in soil science to medical bacteriology during the 1930s–1940s. On the influence of the post-1861 reforms on soil science, see Krupenikov, *Istoria pochvovedeniia*, 155–156.

[95] Trass, *Geobotanika: Istoriia i Sovremennye Tendentsii Razvitiia*, 91.

[96] Richard G. Robbins, *Famine in Russia, 1891–1892: The Imperial Government Responds to a Crisis* (New York: Columbia University Press, 1975), 1.

[97] Ibid., 1–2.

policies in place since the emancipation act of 1861—perhaps exacerbated by I. A. Vyshnegradskii's 1887 reforms—had left the peasant population living in chronic poverty and misery.[98] Especially hard hit were the black earth regions, where peasants responded to the failed harvests with radical measures—devoting more of their fallow and hay fields to crop production, developing a "mania for chopping down trees," and reducing the size of their cattle herds.[99] These actions proved disastrous for Russian agriculture, leading, respectively, to soil exhaustion, the loss of natural windbreaks that lessened the effects of drought, and the loss of much needed fertilizers.[100] Seeking relief for these troubles, the government turned to soil scientists.

The expeditions of 1890s brought together soil science and geobotany just as they were emerging as self-conscious disciplines. Soil science and geographical botany found a common goal and language during the government-sponsored expeditions of the 1880s–1890s. During these expeditions, the founders of Russian ecology participated in a series of expeditions to the agriculturally important chernozem or black earth regions of central and southern Russia.[101] The government commissioned Dokuchaev to organize two "complex expeditions" to study the state of the soil. The first expedition (1882–1886) explored the soils and vegetation formations of the Nizhegorodskaia guberniia, and the second (1888–1894) in the Poltava region near Kiev.[102]

The recognized "father of Russian botany," Andrei Beketov, played a central role in teaching not only Dokuchaev, but also many of the botanists who participated in these expeditions.[103] Through his intellectual guidance and producing of well-trained students, Beketov helped found the highly influential "Dokuchaev scientific school of soil science" and the first Russian school of geobotany.[104] Beketov's students participated in Dokuchaev's expeditions, including those organized while Vinogradskii was his student. Moreover, through his students, Beketov brought his evolutionary concept of natural harmony to the expeditions, and thus to soil science, geobotany, and early Russian ecology. In the late-1870s and early-1880s, he was attempting to reconcile his notion that: "The structure, external appearance and the entire essence of each being is caused by the surrounding conditions and dependence upon these conditions—in short by harmony"—with Darwin's "struggle for

[98] Ibid., 3–8.

[99] Ibid., 8–10.

[100] Ibid., 10.

[101] Ivanov, *Istoriia otechestvennogo pochvovedeniia*, 49.

[102] Ibid.

[103] G. F. Kir'ianov, *Vasilii Vasil'evich Dokuchaev, 1846–1903*, 28.

[104] I. V. Ivanov, *Istoriia otechestvennogo pochvovedeniia:* 45–47, 58, for quote see 105; and Kh. Kh. Trass, *Geobotanika: Istoriia i sovremennye tendentsii razvitiia*, 24. For a discussion of the role of geobotany in soil science see, E. M. Lavrenko, "Znachenie rabot V. V. Dokuchaeva dlia razvitiia russkoi geobotaniki," in A. A. Gigor'ev,. ed., *V. V. Dokuchaev i geografiia* (S.S.S.R.: Izd-vo Akademii Nauk S.S.S.R., 1946), 55–66.

existence."[105] By the 1890s, he had achieved this by replacing Darwin's Malthusian concept of the "struggle for existence" with his own concept of "life competition" and "equilibrium."[106]

Beketov's concept of natural harmony guided the activities of Dokuchaev's expeditions. There his students "examin[ed] thoroughly all the conditions of existence of each being … to discover how these conditions, and the totality of their effects," influenced organic evolution.[107] The geobotanists who participated in Dokuchaev's multifaceted expeditions would come to view the role of vegetation quite differently.[108] Unlike the soil scientists, who focused on soil formations such as the chernozem, the botanists concentrated their attention on vegetation structures such as the forests or steppes, leaving little room for Vinogradskii's study of microbial processes.

At the end of the 1880s, like the soil scientists, geobotanists generally understood their task to be "the study of the dependence between the vegetation and the soil."[109] In 1888, A. N. Krasnov (1862–1914), for example, defined geobotany as the "study of the interdependence of the characteristics of the botanical formations of the plant kingdom, and the history of rock, from which the soil is formed."[110] In his view, geobotany was founded "on soil science in the widest sense of the word."[111] Another geobotanist, who came under the influence of the Dokuchaev school, N. I. Kuznetsov, portrayed the geobotanical method as "characterizing a given formation, that is, a given natural plant association, by clarifying … the biological conditions of combined cohabitation of plants. Of these conditions the most prominent are the soil and climatic conditions."[112] These geobotanists provided the foundation for Russian plant ecology, and the basis for including Vinogradskii's research, in the twentieth century.[113]

---

[105] On the development of Beketov's evolutionary theory, see Daniel P. Todes, Darwin *without Malthus: The Struggle for Existence in Russian Evolutionary Thought* (New York: Oxford University Press, 1989), 50–61. For quotes see, A. N. Beketov, "Garmoniia v prirode," *Russkii vestnik*, Vol. 30, No. 11, 542–543; Op. cit. Todes, *Darwin without Malthus*, 52. On Beketov's critique of Darwin's theory, see Ibid., 56–60; and A. N. Beketov, *Geografiia rastenii: Ocherk ucheniia o rasprostranenii i raspredelenii rastitel'nosti ha zemnoi poverkhnosti* (S.-Peterburg: Tipografiia V. Demakova, 1896), 4–24.

[106] Ibid., 56.

[107] Ibid., 52.

[108] G. I. Dokhman, *Istoriia Geobotaniki v Rossii* (Moskva: Izd-vo Nauka, 1973), 32–33. Collaborating in the expeditions were soil scientists, chemists, meteorologists, agronomists, and botanists.

[109] Ibid., 12.

[110] Ibid. Krasnov was a student of the botanist, Andrei Beketov (see below) and participated in Dokuchaev's 1882 expedition.

[111] Trass, *Geobotanika: Istoriia i Sovremennye Tendentsii Razvitiia*, 33.

[112] Ibid., 14.

[113] Some Western European plant ecologists were familiar with the Russian work, but whether they drew on it in any significant way remains unclear.

Chapter 6
Soil Science and Russian Ecology

Neither Vinogradskii's scientific worldview nor his laboratory research translated easily into the ecological trends of soil science and geobotany. Although he did not participate in the expeditions himself, there are indications that he would have been familiar with their goals, organizational structure, and methodologies.[1] As a new member of the Institute of Experimental Medicine, and a returning student of the St. Petersburg scientific circle, he maintained connections with his previous colleagues. He would have heard about the findings of the Dokuchaev expeditions, which were presented at the meetings of the St. Petersburg Society of Naturalists.[2] He also had contact with the Forestry Institute, seeking their experts out for advice on managing the forests on his Ukrainian estate.[3] Even in close proximity to these scientists, however, he did not adopt their Humboldtian, Darwinian, or ecological language.

Vinogradskii's Contributions

Yet, Vinogradskii supported soil science and geobotany through his general microbiological research at the IEM. From 1892 to 1898, he continued to study the microbial processes important to understanding the nature of the soil. He applied the elective

[1] During his years at St. Petersburg University's botanical department, he had studied with Beketov. Vinogradskii may also have participated in the local excursions Beketov organized. See Chap. 2.

[2] Vinogradskii had been a member of the St. Peterburg Society of Naturalists since 1881. The reports from the expedition appeared in each of the sections of the society and were published in its journal. For a review of the activities of the society between 1888 and 1893, see *Obzor Deiiatel'nosti St Peterburgskogo Obshchestva Estestvoispitatelei*, 1893, Vol. XXV.

[3] Selman Waksman, *Sergei N. Winogradsky: His Life and Work: The Story of a Great Bacteriologist* (New Brunswick: Rutgers University Press, 1953), 32–33.

L. Ackert, *Sergei Vinogradskii and the Cycle of Life: From the Thermodynamics of Life to Ecological Microbiology, 1850-1950*, Archimedes 34,
DOI 10.1007/978-94-007-5198-9_6,

culture method to a series of questions related to understanding the role of autotrophic microbes in the soil. Primarily related to studying nitrogen bacteria physiology, these researches resulted in: first, the confirmation of the biological nature of the nitrogen fixing (1893–1894); second, the isolation of *Clostridium pasteuranium*, the first non-symbiotic nitrogen-fixing bacteria (1895)[4]; and finally, an extensive morphological and physiological taxonomy of the species (1902).[5] Throughout these investigations, he developed the elective culture method into a versatile tool for studying nearly any microbial process occurring in natural soil or water environments.[6]

Throughout this process, he continued his general program of physiological research, by striving to create cultures that would provide environmental conditions favorable for a single vital function. In his own research, Vinogradskii investigated nitrogen fixation—the conversion of free nitrogen gas in the soil into a bound state, thus making it accessible to plants. This phenomenon had previously been related to the presence of organic compounds by agricultural chemists—including Berthelot, Hellriegel and Willfarth, Prazhmovskii, and Schloesing and Laurent. Vinogradskii took greatest issue with Berthelot who had claimed priority not only in determining the chemical characteristics of this phenomenon, but also in isolating the microbe responsible for it. Vinogradskii challenged this claim on the simple basis that Berthelot defined "microbe" so generally that it even included microscopic plants: algae, bacteria, and molds. Ultimately, though, Vinogradskii founded his strong criticism of this work on his own autotrophic investigations. These had bolstered his confidence in, and his commitment to, the idea that microbes are associated with specific transformations of matter in nature. The approach he had developed because of this research—a combination of analytical chemistry skills, macroscopic and microscopic observation, and field studies—gave him the tools to isolate these organisms from the biological chaos of the soil.[7] In the 1890s, however, Russia's soil scientists and geobotanists—its proto-ecologists—struggled to incorporate his work actively into their research.

Vinogradskii also engaged the resources of his department to explore questions related to soil science and geobotany. He assigned topics in soil microbiology to his assistant, Vasilii Omelianskii, and to his praktikanty V. V. Polovtsev and V. A. Fribes. In 1892, Vinogradskii set Fribes to investigate flax retting—the process of soaking flax to soften and separate the fibers by partial decomposition—as a bacteriological

---

[4] Beijerinck had recently discovered bacteria living in symbiosis with leguminous plants in their root nodules. M. W. Beyerinck, "Die Bacterien der Papilionaceen Knollchen," *Botanische Zeitung*, No. 48, 1888, 769–771.

[5] Serge Winogradsky, Sur l'assimilation de l'azote gaseaux de l'atmosphère par les microbes, *Comptes Rendus de Séances de l'Académie des Sciences*, 1893, 1er Semestre, Vol. 116, No. 24, 1385–1388; Idem., *Comptes Rendus hebdomadaires des Seances de l'Académie des Sciences*, Vol. 118, No. 7, 353–355.

[6] Ibid.

[7] Ibid.

problem.[8] This project not only continued Vinogradskii's efforts to improve the elective culture method, it also had practical significance related to the production of textiles. Fribes applied Vinogradskii's elective culture method to isolate the bacteria responsible for decomposing certain kinds of plant fibers.[9] The Imperial Free Economic Society—the same institution that had commissioned Dokuchaev to organize his expeditions—took an interest in this research, and at their annual meeting in 1896, Fribes presented his results "On the Retting of Flax."[10] Through Fribes Vinogradskii maintained a relationship with the Ministry of Agriculture, where Fribes eventually found a position, although he occasionally returned to work with Vinogradskii and Omelianskii.[11]

When Fribes was studying the microbiology of plant decomposition, Vinogradskii assigned Polovtsev to investigate a question central to soil science—whether nitrogen bacteria could decompose organic matter.[12] Vinogradskii organized this project to defend his theory of autotrophism, proving again that nitrogen bacteria could survive only on inorganic matter.

Omelianskii trained with Vinogradskii in microbiology from 1893 to 1899.[13] At the end of that time, Vinogradskii fell seriously ill and returned to Kiev to convalesce, leaving his department in Omelianskii's excellent care. Initially collaborating with Vinogradskii on his nitrogen fixation research, Omelianskii adopted Vinogradskii's methods and cycle of life perspective. In addition, Omelianskii developed a strong sense of the practical applications for microbiology to soil science. After he became the director of the department in 1912—and adopted many of Vinogradskii's other institutional duties—Omelianskii developed a comprehensive program in soil microbiology.

Drawing on his training with Vinogradskii, in 1918 Omelianskii outlined "The Goals of Soil Microbiology."[14] Placing their science firmly in the discipline of soil

---

[8] 1892 Delevoi otchet Otdela Obshchei Mikrobiologii, Tsentral'nyi Gosudarstvennyi Istoricheskii Archiv, St Peterburg, fond 2282, opis' 1, delo 87, fond 2282, opis' 1, delo 31, listy 24 ob.-25 ob.

[9] V. A. Fribes with Serge Winogradsky, "Sur le rouissage de lin et son agent microbien," *Comptes Rendus de l'Académie des Sciences*, 1895, CXXI, 742.

[10] 1896 Delevoi otchet Otdela Obshchei Mikrobiologii, Tsentral'nyi Gosudarstvennyi Istoricheskii Archiv, fond 2282, opis' 1, delo 87, listy 37–37ob.

[11] Fribes visited the laboratory in 1899. 1899–1900 Delevoi otchet Otdela Obshchei Mikrobiologii, Idem., fond 2282, opis' 1, delo 126, listy 146–146 ob.

[12] 1892 Delevoi otchet Otdela Obshchei Mikrobiologii, Idem., fond 2282, opis' 1, delo 31, listy 24 ob.-25 ob.

[13] Omelianskii's publications during the 1890s include: S. Winogradsky and V. Omeliansky, "Ueber den Einfluss der Organischen Substanzen auf die Arbeit der nitrifizierenden Mikroben," *Zentralblatt für Bakteriologie, Parisitenkunde und Infektionskrankheiten, zweite abteilung: Allegemeine, landwirschaftlich-technologische Bakerteriologie, Garungsphysiologie, Pflanzenpathologie und Pflanzenschutz*, May 1899, Vol. V, No. 10, 329–440; W. Omeliansky, "Ueber die Nitrifikation des organischen Stickstoffes," Idem., July 1899, Vol. V, No. 13, 473–490; and V. Omeliansky, "Ueber die Isolierung der Nitrifikationsmikroben aus dem Erdboden," Idem., No. 15, 537–549.

[14] V. L. Omelianskii, "Zadachi pochvennoi mikrobiologii," a draft of his monograph *Mikrobiologiia Pochvy*, Archiv Rossiiskoi Akademii Nauk, Peterburgskoi Filial, fond 892, opis' 1, delo 11, list 1.

science, at the top of his list he recited the mantra of the Dokuchaev school: "The soil as a natural-historical body. Soil formation."[15] Soil microbiologists, Omelianskii taught, would approach this topic by investigating "the disintegration and destruction of minerals and rock under the influence of microorganisms, as an essential factor in soil formation."[16] Using Vinogradskii's elective culture method, soil microbiologists would "reveal the content of microbes in the soil, which provoke defined biochemical transformations," including ammonia fixation, nitrogen fixation, and cellulose decomposition.[17] In a perfect synthesis of Vinogradskii's microbiology and Dokuchaev's soil science, Omelianskii trained soil microbiologists to study how the soil microflora changed under the influence of the various conditions," including the content of organic matter, the absorption ability and structure of the soil, temperature, aeration and humidity, and the distribution and mixture of nutritive materials.[18]

In the 1890s, then, Vinogradskii organized a program of soil microbiology that reflected the interests of soil scientists and geobotanists. They did not integrate his concepts and methods into their discipline, however, until the late-1910s–1920s. That their mutual interest was offset by this disengagement in practice reflects their differences in scientific worldview. This divergence contributed to the independent or parallel development of ecological approaches in soil science and Vinogradskii's microbiology.

Physiological Ecology

In the 1890s, plant ecology crystallized as a self-conscious discipline around the synthesis of natural history—in the form of Humboldtian plant geography—and laboratory-based experimental physiology.[19] This synthesis appeared first in the work of four plant geographers—Eugenius Warming, Oscar Drude, A. F. W. Schimper, and Beketov—whose monographs contributed to the foundation of ecological sciences.[20] Each of these botanists considered physiology to be an important

[15] Ibid.

[16] Ibid.

[17] Ibid.

[18] Ibid. 2–3.

[19] I borrow the metaphor "crystallized" from Robert P. McIntosh, *The Background of Ecology: Concept and Theory* (Cambridge: Cambridge University Press, 1985), 28. For a discussion of the synthesis of Humboldtian plant geography and physiology see: Ibid., 21–27; Cittadino, *Nature as the Laboratory*, 150–157; and Hagan, *An Entangled Bank*, 23–32.

[20] For their discussions of the role of physiology in their plant geography, see: Oscar Drude, *Deutschlands Pflanzengeographie: Ein geographisches Charaterbild der Flora von Deutschland* (Stuttgart: Verlag von J. Engelhorn, 1896), 24–26; and A. F. W. Schimper, "Forward," in *Pflanzen-Geographie auf Physiologischer Grundlage* (Jena: Verlag von Gustav Fischer, 1898), iii–vi; Eugenius Warming, *Lehrbuch der ökologischen Pflanzengeographie: Eine Einfuhrung in die Kenntnis der Pflanzenvereine* (Berlin: Gebruder Borntraeger, 1896); and A. N. Beketov, *Geografiia rastenii: Ocherk' ucheniia o rasprostranenii i raspredelenii rastitel'nosti ha zemnoi poverkhnosti* (S.-Peterburg: Tipografiia V. Demakova, 1896).

part of plant ecology. They limited the role of physiology, however, to the study of whole plants or plant associations, ignoring microorganisms. Considering that they either studied microbiology themselves or had a comprehensive knowledge of the literature, this raises questions about the role of microbiology in early plant ecology. Here we begin to see the distance that separated two developments in the history of ecology—one through evolutionary plant geography and another through thermodynamical microbiology. Drude and Schimper developed their work in terms of orthodox Darwinism and discussed microorganisms only briefly in their works. More informative for understanding why Vinogradskii developed his ecological thinking outside of plant geography is the work of Beketov (who was his teacher) and Warming (whose work played an important role in his *Beggiatoa* research), and especially how they portrayed physiology in their foundational ecological monographs.

Warming's Danish *Plantesamfund* (1895) and its more accessible German translation *Lehrbuch der ökologischen Pflanzengeographie* (1896) is considered, "for all practical purposes, the first textbook of plant ecology."[21] His book attracted the attention of European and American botanists because he emphasized the study of plant communities as assemblages of plants based on their morphology, anatomy, and physiology.[22] Like Beketov, Warming accepted Darwin's concept of natural selection as a satisfactory explanation of plant adaptation, and emphasized equally the direct influence of environmental factors.[23] Warming distinguished ecological plant geography from other botanical sciences such as floristic plant geography by its use of plant physiology "to study the influence of factors … heat, light, nutrition, and water on the form and economy (*Haushaltung*) of plants and plant associations."[24]

Described as such, Warming's science seems amenable to Vinogradskii's research. For Warming, however, the ultimate goal of ecological plant geography was to study the "life form" (*Lebensform*)—a plant's stage of adaptation—and not the physiological types Vinogradskii had investigated.[25] Examining life forms would help ecological plant geographers achieve their most difficult tasks—to explain "*why* species formed together in certain societies and why they possessed certain physiognomies."[26] To answer these questions they would have to investigate the "economy of the plants, their requirements in living conditions … how they use external conditions and how they adapted to them in their external and internal structure, and their physiognomy."[27]

Warming, who had studied microorganisms in the 1870s, included bacteria and microscopic fungi in his approach to plant geography. He treated them, though, not

---

[21] Cittadino, *Nature as the Laboratory*, 147.

[22] Warming, *Lehrbuch der ökologischen Pflanzengeographie*, 3.

[23] For a discussion of how this compared to the German plant ecologists, see Ibid., 148.

[24] Warming, *Lehrbuch der ökologischen Pflanzengeographie*, 2

[25] Ibid., 6.

[26] Ibid., 3.

[27] Ibid.

as plants or associations, but as environmental factors. Like water and temperature fluctuations, for example, they contributed to the mechanical loosening and chemical decomposition of rocks in soil formation.[28] While he used physiology to study the formation of plant associations, he did not apply it to investigate bacteria as part of these societies. In certain areas the natural qualities (*Beschaftenheit*) in the soil or water produced plant-less "wastelands," inhabited "presumably [by] a rich bacterial flora, namely of *Beggiatoa* species."[29] Warming did not incorporate Vinogradskii's sulphur and nitrogen bacteria work into his approach, thus he did not understand those bacteria as physiological types.[30] As he had in his previous investigations, Warming still viewed the sulphur and nitrogen compounds in and around these organisms as independent of their 'life forms.'

In 1896, Beketov made his final contribution to Russian ecology. In a last creative effort, he published *Geografiia Rasteniia* (*Plant Geography*), which he considered the culmination of his attempts to write a comprehensive textbook on botany.[31] Here he developed an approach to ecological plant geography within the framework of his harmony-based, anti-Malthusian evolutionary theory. Drawing on phytogeography (systematics and descriptive botany), paleontology, and physiology, plant geographers try "to explain the action of external physical forces on the forces especially inherent in plants . . . which determine and maintain existing state of plant cover."[32] In plant geography, for Beketov, physiology served as the paramount source for explaining causality, that is, "transmutation—the very process of establishing the present distribution of plants."[33] The study of transmutation, thus, required the knowledge of the geographical position of species.

Beketov viewed ecological plant geography as the study of evolution. By 1895, having excluded Darwin's concept of intraspecific struggle for existence, Beketov argued that the only way to measure speciation was through hybridization studies. To investigate this he proposed a method he called "morphological phytogeography."[34] Plant geographers would "study completely the geographical and topographical distribution of closely related forms, for example, the species of a given genus." They would, in parallel, study these forms' morphologically distinguishing traits, measuring their level of constancy. The most variable traits represented transitional characteristics between two forms. Correlating these forms with their geographical position might reveal speciation in process, whether it occurred by hybridization or assisted by the natural selection of differentiating traits.[35] It was through his concept of natural selection that he introduced physiological methods in ecological plant geography, and, thus, made it accessible to Vinogradskii's research.

[28] Ibid., 41–42.

[29] Ibid., 140. See Chap. 2 for a discussion of Warming's sulphur bacteria research.

[30] Warming discussed nitrogen bacteria in, Ibid., 75–76.

[31] He had published his first botany textbook in 1883. Beketov, *Geografiia rastenii*, 1.

[32] Ibid., 3–4.

[33] Ibid., 4.

[34] Ibid., 12.

[35] Ibid.

Considering Darwin's Malthusian concept of 'struggle for survival' a false violation of his own view of natural harmony, Beketov had developed an alternative mechanism he called "life competition" (*zhizennoe sostiazanie*).[36] By life competition he meant an intraspecific struggle for resources that never led to the elimination of a species, but rather to new stages of equilibrium (or harmony). Searching in all of nature, Beketov found only fluctuations and equilibrium, in spite of the eternal antagonism of forces. This led him to conclude that life competition was itself an antagonistic force that led to "a struggle for equilibrium."[37] Of the circumstances that determined life competition: (1) limited space, which determined the amount of resources for existence; (2) the geometric propagation of organisms, physiology would be used to study, (3) the relationship of organisms to the external conditions. Here we see that this relationship—one central to Vinogradskii's scientific worldview—was buried within a complex evolutionary system couched in Darwinian terms that lay outside of Vinogradskii's thermodynamic vision of life.

Beketov Without Darwin: Vinogradskii's Concept of the Cycle of Life

In the twentieth century soil scientists, geobotanists, and microbiologists came to view Vinogradskii's work in the 1880s–1890s as setting the foundation of ecological microbiology. During the 1890s, however, his work had little impact on the development of other ecological sciences. Soil scientists and geobotanists did incorporate the concept of autotrophism into their theoretical framework, but found little practical use for his laboratory research on microorganisms. Being a plant physiologist, Vinogradskii might have contributed to the programs being developed by ecological plant geographers, yet his laboratory research did not support their plant community focus, and geographical and cartographic methodologies.

The distance between Vinogradskii and other early ecologists grew from a difference in perspective. The plant geographers and soil scientists studied the distribution of whole organisms and their societies across geographical spaces and geological time, whereas Vinogradskii investigated microbial function in microscopic landscapes. Where they developed their research programs in terms of evolutionary theory, Vinogradskii paid little attention to it. Instead, he conducted his physiological, microbiological, and soil science research from the cycle of life perspective—a holistic, thermodynamical approach to the study of nature.

Vinogradskii's developing ecological perspective contrasted with those of the plant geographers most starkly in his public presentations during the 1890s. In 1894, at the Russian botanist Kliment Timiriazev's request, Vinogradskii presented his

[36] Ibid., 16–17. For a discussion of the significance of Beketov's concept of "life competition" for the reception of Darwinian theory in Russia, see Todes, *Darwin without Malthus*, 58–60.

[37] Ibid., 18.

first talk to a Russian lay audience. Vinogradskii unshackled his vision of nature from the details of his laboratory data and introduced his cycle of life vision of nature to the broader public. At a Moscow conference—most likely of the Society of Naturalists—Vinogradskii discussed "The Nitrogen Cycle in Nature."[38] In this speech, he "connected together all of [his] microbes" in hopes of stirring his audience's imagination with the great things small microbes can accomplish.[39] While the details of this talk are lost, he probably discussed the concept of the cycle of life—a theme that would have resonated strongly with his colleagues of this period.

Vinogradskii expressed a very different view of ecological relationships than those being described by the plant geographers in 1890s. He described his vision in a popular lecture at the IEM "On the Role of Microbes in the General Cycle of Life."[40] He told his audience that he was somewhat apprehensive about discussing microbes—recently physicians had been stressing the dangers and harms of bacteria and others had observed the industrial or agricultural utility of microbes. He reached out, instead, to those "thinking persons who were ready to fathom the general significance of new facts within their worldview, to apply to them that logic which one is accustomed to find in nature's phenomena."[41] He chose to summarize the significance of the microbes in nature according to his own worldview, according to which "the motion of matter is caused by the phenomena of life in all their totality." He observed that living beings draw the material they need to build their bodies—carbon, nitrogen, hydrogen, and oxygen—from the reservoir of mineral nature. This posed a problem, though. If the accumulation of these minerals continued endlessly in one direction without a reverse process, then nature's reserves would sooner or later become depleted.[42]

He sought an escape from this dilemma not in Humboldtian plant geography or Darwinian evolution, but from Pasteurian microbiology. The answer for Vinogradskii was the "circulation of matter," without which "the regular and prolonged existence of the organized world would be inconceivable."[43] Referring his audience to the work of the French chemists Dumas and Boussingault, in which "all that the air gives to the plants, they give up to the animals; and the animals then return it to the air—an eternal circle, in which life revolves, yet matter only changes place," Vinogradskii noted both its ingeniousness, and its incompleteness.[44] Missing from this simple and graceful system, he said, was both an understanding that a condition of equilibrium existed between the reserves of dead nature and the world of the

[38] Letter from Vinogradskii to Elie Metchnikov dated 9 February 1894, *Bor'ba za Nauku v Tsarskoi Rossii*, 167.

[39] Ibid.

[40] This speech was later published as S. N. Vinogradskii, "O Roli Mikrobov v Obshchem Krugovorote Zhizni," *Arkhiv Biologicheskix Nauk*, Vol. 7, No. 3, 1897, 1–27; for the quote see 26–27.

[41] Ibid., 5.

[42] Ibid., 5–6.

[43] Ibid., 6.

[44] Ibid., 7–8.

higher organisms, and also a mechanism for transforming dead organisms into single inorganic compounds. Here he recited Pasteur's well-known observation that: "life presides over the work of death in all of its phases."

Vinogradskii impressed upon his audience the magnitude and purposefulness of this microbial work. There were billions of kilograms of organic matter to be decomposed at any time. Laboratory research had shown that powerful acids and very high temperatures were needed to break down organic bodies, and this succeeded only incompletely. Yet microbes were able to accomplish this "easily and unnoticeably."[45] To decompose the great number of organic materials in varied natural conditions microbes must possess specialized, diverse, and energetic reagents. These "agents of decay" were not, for him, mere chemical analyzers; they were living, growing, and breeding beings, and as such, the material they decomposed was their food.

Using physiology, microbiologists would determine the relationship between microbial species and decomposition processes, according to the principle of the division of labor. Where plant geographers strove to map plant societies, Vinogradskii envisioned compiling tables (reminiscent of Dmitrii Mendeleev's table of elements perhaps), in which each organic body (in column one) would be associated with a name or number of a microbe that could decompose that body (in column two), and the product of that decomposition (in column three). It was clear to him that the microbial world was organized by the principle of the division of labor—a close correspondence had been discovered to exist between a microbe's physiological characteristics and the molecular structure of the organic body it could decompose.[46] Accordingly, the function of microbes in nature was specialized—"each job had its own specialist"—but those functions were not strictly demarcated. In nature, for Vinogradskii, there were no sharp boundaries—each function was part of an entire series of imperceptible transitions. This overlapping in function created competition between species for resources.

He explained the purpose of this competition, not in terms of Darwinian theory, but as a physiological process of production or "cultivation."[47] Classifying microbes according to the bodies they could decompose revealed groupings of species that could eat the same material, for example, sugar. Each species, however, worked more energetically in some conditions than in others. Laboratory investigations had shown that when several species were combined in a certain environment, the most energetic species would "take possession" of the culture. He described this as a "struggle for existence" between species with similar functions. Physiology is central here—the especially intense character of microbial nutrition causes an intense struggle. In nature, this struggle maintains microbial activity at its highest level, producing species of very high energy. Vinogradskii, moreover, had observed this in his own experiments, during which he succeeded in "cultivating" new races of energetic microbes using "non-artificial methods."[48]

---

[45] Ibid., 10.

[46] Ibid., 14.

[47] Ibid., 16.

[48] Ibid.

In his talk, he had yet to address how this view of microbes solved the problem of a unidirectional flow of material from the organic to the inorganic realms. Drawing on his nitrification research, he now turned to this question. His Zurich research had shown that specific microbes performed "a final act of regressive metamorphosis" by chemically "analyzing" nitrogen from ammonia. As he described it, they had now discovered the very origin of the "progressive metamorphosis of matter."[49] Subsequently, through his nitrogen fixation work at the IEM, he had discovered *Clostridium pasteuranium*—the first organism that could begin this progressive metamorphosis and "synthesize" nitrogen into compounds that plants could assimilate.

With the mystery of the nitrogen cycle solved, Vinogradskii sketched the general role of microbes in nature:

> Microbes are the main (*glavnyi*) agents called forth by life and are necessary for the lawful operation the cycle of life (*pravil'noi smeny zhiznei krugovorota veshchestv*). They are the living bearers of infinitely varied reactives, and one can even say, they are the reactives incarnate, without which many of the necessary process of that cycle would be inconceivable. It is clear to us that only the fundamental qualities of living beings—the ability of propagation, spreading, adaptation (*prisposobleniia*), and heredity—can provide these processes their necessary plasticity, spontaneity, and inevitability.[50]

Sounding the same note that reverberated through his scientific investigations, he described the cycle of life as "a single entity, a single huge organism that borrows its elements from the reserves of inorganic nature, purposively manages all the process of its progressive and regressive metamorphoses and, finally, which returns again all that was borrowed to dead nature."[51] In this vision of the cycle of life, Vinogradskii delegated the essential role to microorganisms—only through their agency did matter circulate between nature's two realms.

In his talk, Vinogradskii described the worldview that guided his laboratory investigations in his apprenticeship, in Strassburg, Zurich, and at the IEM. He would be committed to this vision of a cycle of life driven by the purposeful physiological characteristics of microbes for the duration of his career. In 1899, Vinogradskii again extrapolated from the struggle for resources he observed through his microscope to a grand nitrogen cycle operating on the entire planet.[52] Synthesizing his observations on the "great sensitivity" of nitrogen bacteria to organic materials, he enumerated the series of nature's physiological stages that occurred in the soil: ammonification, nitrification, and denitrification.[53] He imagined nature as a living organism in which microbes responded sensitively to their conditions of existence—living in a tight competition for materials—thus performing the vital functions of that super-organism.

---

[49] Ibid., 21–22.

[50] Ibid., 27.

[51] Ibid.

[52] Serge Winogradsky, "L'Influence des Substances Organiques sur le Travail des Microbes Nitrificateurs," in *Microbiologie du Sol: Problèmes et Méthodes—Cinquante Ans de Recherches* (Paris: Masson et Cie, 1949), 234.

[53] Ibid., 235–238.

Comparing Vinogradskii's vision of nature and the experimental physiological approach he developed to explore it, with those of the plant geographers sets in sharp relief the features of their different ecological perspectives. As with the soil scientists and geobotanists, these distinctions reflect the independent development of Vinogradskii as an ecological thinker.

Scientific Forestry: Vinogradskii Retires to Gorodok

In St. Petersburg, Vinogradskii confined his interest in soil science to the laboratory investigations of nitrification and related microbial soil processes. On his Gorodok estates, however, he applied his knowledge to practical questions in agricultural. Developing nephritis in 1898–1899, gave him a troubling and unexpected opportunity to explore more fully a career he had abandoned for science—that of a *pomeshchik* or large-scale landowner. He could not have returned, however, to the kind of life lived by his father—Vinogradskii had been transformed by his scientific education and career. When not busy with his institutional, society, or research duties in St. Petersburg, he devoted his time to "scientific farming" on his large estate, which included nearly 5,000 acres of old growth forest, horse stables, an extensive beet-sugar factory, and several flour mills.[54] Drawing on the expertise he had developed in St. Petersburg and the assistance of experts from the Forestry Institute of Kiev, he decided which trees to cull and which to leave, and established an experimental nursery. In addition, he organized a model dairy farm and an orchard.[55] These activities suggest that Vinogradskii's interests in farming had shifted from that of a *pomeshchik*, or landlord, to those of an agriculturalist and soil scientist.

In his mid-60s, Vinogradskii no doubt planned on living out his life as a *pomeshchik*. In 1918, however, the upheaval caused by the Russian Civil War—especially destructive in the Kiev area—forced him and his family to flee for their lives. Brought on by the Bolsheviks he came to despise, this hardship awakened him from his latent life and opened a new chapter in his career. To his enemies, then, he may owe his fame for bringing ecology to microbiology.[56]

---

[54] Selman Waksman, *Sergei N. Winogradsky: His Life and Work: The Story of a Great Bacteriologist* (New Brunswick Rutgers University Press, 1953), 32–33.

[55] Ibid.

[56] From a Russian outline of his professional activities "Copie d'un formulaire allemand sous l'occupation," written when he was 86 years old (1942); Serge Winogradsky Papers, Service des Archives de l'Institut Pasteur, Box WIN 2, Folder Serge Winogradsky, Correspondance, France; see 1, 7.

Part IV
French Agriculture

Chapter 7
The Master of Brie-Compte-Robert and His "Direct Method:" Translating the Cycle of Life into Ecology

> *It is customary to recognize as a department of biology the science called ecology. The name is abominable; the science is one of great beauty. Ecology concerns the relations between living things and their environment. . . . But before Pasteur's day no man could have a clear idea of many of the interrelations between organisms.*[1]

After the storm of the Russian Civil War, Vinogradskii quietly reassessed his life. He had lost all of his material possessions except for a Swiss chalet and a small cache of money—his most valuable remaining asset was his scientific credibility. Drawing on this latter, he resurrected his career, found a new institution, and engaged a new movement in science. Working for a new patron—the Pasteur Institute—he now applied his three decades of experience in microbial physiology and soil science to practical use in agricultural microbiology. Specifically, he adapted the "elective culture" method he had developed in the 1890s to help microbiologists apply their soil science research to practical questions in agriculture. He thus assembled a "direct, ecological method" that he thought would enable microbiologists to study microbes in the complexity of their natural environments.

Attracted to the language of ecology that was emerging as a powerful force in the scientific community in the early twentieth century, he translated his cycle of life worldview, and the experimental program he had developed to explore it, into an ecological research program. By the 1930s, Vinogradskii considered not only his current investigations, but also all of his previous work to be ecology. His first investigations

[1] Lawrence T. Henderson, "Life and Services of Louis Pasteur," *Proceedings of the American Philosophical Society*, Vol. 62 (1923), iii–xiii. My thanks to Sharon Kingsland for pointing out Henderson's comments on Pasteur's relationship to the history of ecology.

L. Ackert, *Sergei Vinogradskii and the Cycle of Life: From the Thermodynamics of Life to Ecological Microbiology, 1850-1950*, Archimedes 34,

DOI 10.1007/978-94-007-5198-9_7,

on *Mycoderma vini*, his discovery of chemosynthesis (the keystone in the cycle of life), and his extensive investigations of the nitrogen cycle—all, he now wrote, expressed the ideals of twentieth century ecology. Remaking himself as an ecologist, he transformed the biological approach he had developed over four decades of physiological and microbiological investigations into a novel research program that excited a new generation of microbiologist, medical bacteriologists, and soil scientists.[2] For Vinogradskii—a Pasteurian microbiologist steeped in Russian soil science—an *ecological* approach meant investigating a microbe's role in the cycle of life physiologically, that is, exploring its nutrition and respiration as a natural process mediated by the plenum of environmental conditions.

Thus inscribed in Vinogradskii's ecological method, the cycle of life played a role in three disciplines that were fundamentally transformed in the 1920s–1930s: ecology, soil science, and soil microbiology. From a variety of platforms—articles in professional scientific journals and applied science reports, speeches at conferences, his correspondence with an international network, and mentoring at the Pasteur Institute—he simultaneously promoted his own career, his methods, and the development of new scientific fields. Through these efforts, his vision of an ecological soil microbiology radiated through a community of agricultural researchers spread around the world in agricultural experiment stations and bacteriological and microbiological institutes. By adopting his methods these scientists translated Vinogradskii's cycle of life concept into ongoing research programs, making it available for new generations of scientists.

Vinogradskii Comes to Brie-Comte-Robert: The Resurrection of a Career

Vinogradskii found his way to Brie-Comte-Robert and the Pasteur Institute by following his heart. As seen in the last chapter, he endured a tumultuous 2 year period fleeing the dangers of the Russian Civil War. During this "whirlwind tour" of Europe, recognizing that two of his professions—agriculturalist and sugar beet industrialist—were no longer viable career options, he revived his identity as a microbiologist.[3] Although he did write one scientific article (on iron bacteria), he had spent most of his energies writing political articles for anti-Bolshevik

[2] Here I use the idea of a 'new generation' to mean a group of scientists of all ages and at all stages of career, who by the early twentieth century were already primed with a sense of the ecological and who were engaged in organizing new ecological disciplines.

[3] Vinogradskii felt that his work had been misappropriated and misunderstood. In his own words, "he set out to recapture a topic he felt was his own and put it back on the right track." Little did he suspect at this point that this would lead him to ecology; Serge Winogradsky, "Sur la méthode directe dans l'étude microbiologique du sol," *Comptes Rendus Hebdomadières des Séances de l'Académie des Sciences, Paris*, Tome 167, Juillet-Décembre 1923, 1001–1004.

newspapers.[4] The obvious futility of his writing campaign led him to concentrate anew on his scientific possibilities. At the University of Belgrade, he spent his time catching up on the microbiological literature. The scanty resources of the University of Belgrade included a complete run of the *Zentralblatt fur Bakteriologie* between 1895 and 1920. He surveyed this journal to re-familiarize himself with the state of affairs in microbiology, and to mine it for new projects. He would initiate several of these projects at his laboratory in Brie-Compte-Robert over the next 30 years. The notebook he left behind reveals that his reading of the *Zentralblatt* awakened his interest in the ecology applications of microbiology.[5]

Dissatisfied with the poor conditions at the University of Belgrade, he sought a place where he could conduct research. Drawing on his many international contacts, he visited Emil Roux, director of the Pasteur Institute in Paris, and requested a position there. Having recently received a large donation from a wealthy patron, Roux used it to create a position for Vinogradskii as director of a new laboratory of agricultural microbiology. Benefiting from Roux's timely windfall, Vinogradskii organized a science estate (which included a small castle, a gardener's house, and a two-story laboratory) in Brie-Comte-Robert, a small town outside of Paris.[6] Here, Vinogradskii launched a campaign to organize a new research direction at the Pasteur Institute and in general microbiology—that of agricultural microbiology.[7]

[4] As his own life in science seemed at its end, Vinogradskii launched a campaign to create a new place for scientists in society. Supposedly secure behind his pseudonym "Starnat" (for Old Nat[uralist]), he explained his agenda in a critique of the Bolshevik government. If anyone should experiment with society, Vinogradskii argued, it should be well-trained scientists and especially "naturalists." He accused the Bolsheviks of appropriating the language of science without either properly understanding its methods and requirements or dismissing them completely. Seeing no "experiment" in their brutal methods of social reconstruction, Vinogradskii criticized their merely rhetorical use of "experiment." If Lenin, Bukharin, and Trotsky desired to conduct a social experiment, Vinogradskii wrote under the pseudonym "Starnat" for "Old Naturalist," they should do it correctly—they should rely on the naturalist's method of choice, using direct observation of social phenomena in controlled environments. Not wanting to merely criticize, he proposed his own political platform—his "Naturalist's party" offered to manage society by applying the scientific method properly. The frustration of his failed political critiques in the political chaos of this period and his increasing dissatisfaction with his professional situation in Belgrade drove him to search for new employment. Within the framework of this political attack, he applied his cycle of life concept to human society—just as microbes have specific roles to play in nature's economy, so do certain disciplines have in human society. Through these articles he expressed his anger at the Bolshevik party for its heavy-handed revolutionary tactics and indignantly rejected Bolshevik rhetoric about the "Soviet Experiment." Sergei Vinogradskii, "Experimental Socialism," a draft of his article later published in "Iuzhnoe Slovo" in 1919. See Serge Winogradsky Papers, Service des Archives de l'Institut Pasteur, Box WIN GF.

[5] Sergei Vinogradskii, *Bibliographie du Zentrbl. F. Bakt., 1895–1920*, Serge Winogradsky Papers, Service des Archives de l'Institut Pasteur, WIN 3, Box 3, "Bibliographies."

[6] Letter to Emil Roux, 15 February 1922, Archives de l'Institut Pasteur, Serge Winogradsky Papers, Emil Roux Folder.

[7] Vinogradskii's experience at the Imperial Institute of Medicine had taught him that basic research in agricultural microbiology could be a part of a medical research institute. This lesson informed his plans for the laboratory at Brie-Comte-Robert—the greatest goal he could achieve there, he felt, would be to add a new dimension to the Pasteur Institute. With its fame in pathology would stand Vinogradskii's own science, soil microbiology.

At Brie-Comte-Robert, he investigated the role of microbes in the soil using a method that, he contended, provided microbiologists with a "direct" connection to ongoing soil processes. This method combined techniques he had applied since his apprenticeship. He now offered his direct method as a solution to what he viewed as a serious problem in soil microbiology. His retooling of the direct method as an ecological approach represented a complex gesture combining a flexible rhetorical strategy and rigorous scientific research program. This reformulation and research, primarily conducted on soil nitrogen bacteria, must be interpreted within the context of the biographical and autobiographical writing with which he was occupied during the last 20 years of his life. The demands of Vinogradskii's new institutional position to develop practical methods in soil microbiology inspired him to repackage his previous techniques into a novel method, a method that translated the cycle of life into ecological microbiology.

The Direct Method in 1923: Its First Explication

On November 19, 1923, 1 year after opening the Brie-Comte-Robert laboratory, Vinogradskii complained in the Academy of Science's journal, the *Comptes Rendus*, about the fragmentary and imprecise notions that microbiologists possessed concerning microbial phenomena as they occurred in the soil.[8] During the previous 30 years, scientists had succeeded in isolating a number of soil microbes and in reproducing in pure culture most soil processes of interest to agronomists.[9] The great majority of these investigations, Vinogradskii lamented, focused on single species that had been maintained in laboratory collections isolated from its natural environmental conditions. For him, studying microbes cultivated on artificial milieu in conditions of pure culture revealed little about the "the wild [*sauvage*] existence of some species."[10] Cultivating a microbial species in pure culture—sheltered from its "vital and hypernutritional competition [*concurrance*]"—produced rather quickly, he thought, a *plante de culture*—a new race very different from its "prototype."[11]

Drawing on the distinction between wild and cultivated races of yeast introduced long ago in the fermentation laboratories, he argued that agricultural microbiologists should avoid the inauthentic "ancient cultures of the laboratory."[12] No matter how detailed these physiological investigations appeared, by using cultivated microbes in pure culture they failed to address the proper goal of soil science—"studying the

[8] Serge Winogradsky, "Sur la méthode directe dans l'étude microbiologique du sol," 1001–1004. It is significant that this article appeared under the heading *Microbiologie agricole*—another sign that Vinogradskii was now stressing the practical, agronomic aspects of his research.

[9] Soil processes included nitrification, nitrogen fixation, ammoniafication to name only a few.

[10] Ibid. Recall that Vinogradskii distinguished in a similar way between wild nature and artificial cultures as far back as his 1884–1888 sulphur and iron bacteria investigations in Strassburg.

[11] Ibid., 1002.

[12] Ibid.

species that populate the different soils and their functions as they manifested in nature."[13] He set out to correct the situation by conducting a battery of "direct experiments."[14]

At the turn of the century, scientists prided themselves on having direct access to nature's secrets through their laboratory methods.[15] Medical bacteriologists were confident that Robert Koch's pure cultures, four postulates, and refined observational techniques had put their science on a scientific foundation by providing a method for direct investigations of bacteria. For Vinogradskii, agricultural microbiology could not be reformed by using pure cultures. This classical method "could not produce any serious progress in the microbiology of the soil" because it was based on experiments conducted with pure cultures in artificial milieus. Moreover, his literature survey had revealed no alternatives. The methods investigators had proposed for placing "our branch of microbiology," as Vinogradskii called it, at the service of "soil science" were not sufficiently direct.[16] He appealed for adopting a new, more direct method, even if it would initially be "less perfect," than the current technique.[17]

The direct method to which Vinogradskii referred—the one that would bring microbiology to soil science—was his own. With his method soil microbiologists could study microbial activity in conditions as natural as possible.[18] Describing this method in a combination of biological and physical language, he explained that his direct approach consisted of studying the "biological relationships" [*rapports*] that reign between soil microorganisms. He focused on these relationships because to him they represented the processes by which microbes regulated the chemical transformations central to the cycle of life: nitrification, denitrification, and ammoniafication. Envisioning these transformations as "an incessant struggle [between microbes] to appropriate their energy sources," Vinogradskii was attempting through his direct method to investigate natural microbial action in controlled laboratory conditions.[19]

Vinogradskii, then, again introduced the complexity of nature into the laboratory. No longer did the standard pure culture of the classical method provide the principal

---

[13] Ibid., 1001–1002.

[14] Ibid. The notion of using "direct" observations as a rhetorical maneuver has a long history that is related to the rise of sensationalist natural philosophers. It does seem to have a peculiar significance in the history of microbiology, in which so much of the research is conducted through the microscope.

[15] This is an important distinction in the story at the agricultural experiment stations in New England—they had adopted Koch's methods, and were challenged by Vinogradskii's non-Kochian, non-Darwinian approach.

[16] Here Vinogradskii referred to the genre of work inspired by Remy-Lohnis, which "consisted of drowning a little earth in a quantum of liquid, at a rate of 10 per 100, and to follow the processes which developed in the phials [shallow vessels] and their relations with the fertility of the soils employed for the inoculations." Ibid., 1003. It is ironic that Vinogradskii himself would face the same criticism he had of this method (that one could not find the truth about soil in a "pond") from his Soviet colleague, Kholodnyi.

[17] Ibid., 1002–1003.

[18] Ibid.

[19] Ibid. Vinogradskii had found in his recent literature survey "one serious attempt" to study microbes in their natural milieu. Although he does not reveal this source in this document he is most likely referring to H. J. Conn's 1919 method.

environment for his cultures. Now he used "fresh soil, sifted and reduced to a convenient degree of humidity and density."[20] Serving as the basis for all his experiments, this fresh soil also provided the experimental organisms: "germs, already present there, were made to proliferate."[21] By adding this natural soil to various artificial milieus containing chemical bodies chosen specifically for their nutritive abilities—raw materials of animal and vegetable origin—he created a series of "auxiliary cultures."[22] Using direct microscopic examination of these auxiliary cultures, he strove to obtain a complete picture of what he called "released processes" [*processus déclenchè*] or the microbe's "biological reaction."[23] It is clear in reading Vinogradskii's writings from this period that he is searching for an appropriate language to describe the complexities he witnessed in his research, both of primary and secondary sources, and we will see that 'ecological, ecology' would serve his purposes best.

He applied his direct method to investigate the soil; testing the changes from a soil's "microbiological state," as he called it, to its biological reaction.[24] He defined the "microbiological state" as the quality and quantity of active microbial cells in the soil.[25] Its "biological reaction" consisted of the changes provoked in the soil by the addition of energizing substances. The reaction appeared as an "enormous proliferation of a species, or of a small group of the best adapted species."[26] The biological reaction depends, he clarified, not only on the quantity of the energizing substance, but also on the accompanying conditions of the soil (nitrogen level, humidity, and level of aeration). One could study the effect of these conditions on the biological reaction by interrupting the experiment (which often displayed a rapid succession of reactions) at just the right time. By comparing the chemical composition of the culture at that exact, frozen moment with that of the control soil, he could make claims about the microbial population as it lived in its natural conditions. In addition, this method allowed him to meet his commitments to natural history. As the experiment matured, he could single out the single "victorious [*vainqueur*] microbe" responsible for the single biochemical process he was investigating in the complexity of its natural environment.[27]

[20] Ibid., 1003.

[21] Ibid., 1004.

[22] Ibid.

[23] Ibid. The types of solid milieu determined the success of the experiment—they must correspond in composition as close as possible to the soil-environment [*milieu-terre*]. That is, he explained, they must contain the same energizing substance and the same proportion of easily assimilated nitrogen. Previously used "*standard* milieus" were unsuitable because they did not manifest the desired "specialized functions" and were to be avoided.

[24] Ibid., 1004.

[25] Here he equated the microbiological state (*etat microbiologique*) with the biological state.

[26] Ibid., 1004.

[27] Ibid. In his description of the correct match between the control soil and artificial milieus, Vinogradskii revealed his view of the natural state of microbes in the soil. The preferred soil was of ancient culture, that is, it had not been influenced in any way in the recent past and had not received any manure for years, and did not contain any vegetation. This soil—being in "a state of relative equilibrium" and populated "by a stable minimum of microbes"—would react intensively to the energizing substances.

Having transformed the natural microbiological state of the soil into a series of reactions in gelatin plates, Vinogradskii drew upon an increasingly popular laboratory tool—photography, and precisely drawn colored plates. To portray the static information in his gelatin plates as a dynamic process in natural soil conditions, he captured these reactions in a series of pictures. By placing these images next to one another he hoped to show change over time and thus to capture the wildness of the natural soil processes. Moreover, these photographs offered a new rhetorical device for supporting his conclusions. To reconstruct these processes photographically, he needed first to determine the "normal" condition of the soil. He brought these normal soils from their natural location into the laboratory, turning them into control soils (*terre temoin*). These he photographed to capture what he called the "biological content of the soil."[28] He observed, however, that these images failed to capture the dormant microbial species that were hibernating in the form of spores. Thus, in order to understand the dynamics occurring between these microbial species one needed to disrupt the soil's "biological equilibrium" by introducing new environmental conditions—his auxiliary cultures.

To control this "biological reaction," as he called it, he adapted his elective culture method of the 1890s to induce microbial competition for particular nutritive solutions. Then, by taking a picture of this new state, using microphotography or drawings, he contended that this approach provided microbiologists direct access to the "microscopic landscapes" (*Paysages microbiens*) of the soil.[29] This method allowed him to observe the influence of the "totality of natural forces on the make-up of a microbial population," and, also, to investigate a microbe's vital function, that is, its role in the cycle of life. The innovative aspect of the direct method in relation to his earlier methods, Vinogradskii contended, was that it did not rely on pure cultures—microbial action was directly observed in his microscopic landscapes.

To make microphotography work, Vinogradskii drew on the dye technologies being perfected by H. Joel Conn, an American bacteriologist. Vinogradskii borrowed more from Conn, however, than his staining technique, he attributed the direct microscopic method for soil science to the method Conn advanced in his 1918 "The Microscopic Study of Bacteria and Fungi in Soil." In this report, Conn discussed a method he devised to stain dried soil infusions, which allowed for "the direct microscopic study of the bacteria in the soil."[30] Here he drew on the method developed by his colleagues at Geneva, Robert S. Breed and James D. Brew for counting bacteria in milk.

Exploring the history of Conn's direct method helps to understand the impact Vinogradskii's contributions—and subsequent modifications of Conn's method—had on soil bacteriology and related fields.[31] Conn developed the direct microscopic

[28] He used soils that had not been fertilized for at least 3 years, sifting them and then dyeing them using a procedure developed by the American bacteriologist H. Joel Conn.

[29] Here Vinogradskii is using a notion of landscapes very similar to that developed by the Dokuchaev school of Russian soil science. See Chap. 5.

[30] Harold Joel Conn, "The Microscopic Study of the Bacteria and Fungi in Soil," *New York Agricultural Experiment Station Technical Bulletin*, No. 64, January 1918, 1–20, See 3.

method when trying to convert Koch's postulates to use in soil bacteriology. When Vinogradskii read Conn's work (7 years after its publication) he must have recognized a like-minded scientist. Conn's ideas about studying microorganisms in their natural habitat clearly resonated with Vinogradskii's own general approach to soil science. Conn had his own perspective—one informed by current trends in bacteriology and the work of his father, H. W. Conn.

The younger Conn—influenced by the milk bacteria research being conducted by his father and two assistant bacteriologists at the New York Agricultural Experiment Station in Geneva, Robert S. Breed and James D. Brew—translated Koch's postulates for general bacteriology into a program of research in soil bacteriology.[32] Conn thought the "case of bacterial activities in soil analogous [to that of] proving a given organism to be the cause of a given disease."[33] It was not sufficient to assume a causal relationship between the presence of a certain organism—in manured soil, for example—and the processes that occurred there—such as the decomposition of the manure." Furthermore, observations of phenomena in laboratory media do not prove the existence of those phenomena in the soil. He admitted that although soil bacteriologists may have recognized this fact, none (with one exception) had ever followed rules as strict as Koch's postulates "in establishing the agency of bacteria in any soil activity."[34] By way of supporting evidence, Conn cited the research on the bacteria living in legume nodules—thinking no doubt of the well-known work of Beijerinck and Vinogradskii on nitrifying organisms. Even here, however, Conn did not accept this work as "complete proof" of his case.

This lack of definitive proof led Conn to formulate strict rules for investigating "the activities of soil microorganisms."[35] Conn thought it necessary to modify Koch's postulates in particular ways, because soil microorganisms operated in quite different conditions than pathogenic bacteria. For example, Koch's first postulate—the organism must be shown to be present in abundance in animals suffering from the disease in question—could be applied to bacterial activities in the soil. As with pathogenic bacteria it was equally necessary to link the abundance of an organism with a particular biological activity in the soil; however, one needed to consider two additional concerns. First, to assert that the organism is the causal agent one needed to show that it was more abundant in a soil where the biological activity occurred than in similar soil in which the activity was not taking place. Second, it was necessary to show that the organism is in an *active* form.[36] Conn viewed these two new steps

[31] For a recent discussion of the American context during this period see Eric Kupferberg, *The Expertise of Germs: Practice, Language, and Authority in American Bacteriology, 1899–1924* (Harvard University, Ph. D. Dissertation, 2001).

[32] Conn, 253.

[33] Ibid.

[34] Conn, 252–253.

[35] Ibid., 253.

[36] Ibid. This is necessary because at three groups of soil microorganisms have inactive and well as active forms, including the protozoa, molds, and spore-bearing bacteria.

as a "special stringency" required to overcome the difficulty in applying Koch's postulates to soil conditions.[37] He translated Koch's first postulate to address these concerns: "the organisms in question must be shown to be present in active form when the chemical transformation is taking place; and must also be shown to be present in larger numbers in such soil than in similar soil in which the chemical change is not taking place."[38] Koch's second postulate—that the organism be isolated and cultivated in pure culture—could be (and was being currently) applied without modification to soil conditions.

Conn's interest in a new method stemmed from his concern that it would be impossible to meet the demands of Koch's third postulate—to isolate and cultivate the organisms in pure culture—with soil bacteria.[39] While it was possible to grow an organism in sterile soil and study its activity under such conditions, such common tests did not furnish conclusive proof. Drawing a distinction between sterilized and natural soils, Conn thought that these different soils challenged the validity of Koch's postulate for soil science—the activities that occurred in pure cultures would most likely be very different from those in mixed cultures.[40] Complete proof would consist in growing the organism in unsterilized soil, observing whether the microbial process under investigation occurred, and if it did determine whether the organisms populated the culture in large numbers. Conn regretted that this procedure was "generally impossible"—the organism would not grow vigorously in a natural soil "already stocked with a bacterial flora of its own."[41] When, moreover, one did achieve a vigorous growth it still led to confused interpretations because—"in distinct contrast to the specific agency of microorganisms in disease—the same chemical transformation in the soil may be caused by distinctly different organisms."[42]

Sterilized soil, then, provided the only alternative for exploring these questions. It was better that natural soil, which contained multitudinous flora, and also preferable to "any pure culture medium." A reliance on these latter, he thought, had created much of the past confusion regarding the activities of soil bacteria because "pure cultures never occur naturally in the soil."[43] Again, the impracticability of using unsterilized soil made this postulate of little value for soil bacteriology. One could obtain these facts, as Conn called them, using existing cultural methods. Serious errors could occur, however, because "organisms that are naturally inactive may become active under cultural conditions, while under similar conditions naturally

[37] Ibid., 253–254.

[38] Ibid.

[39] Ibid., 254.

[40] Ibid.

[41] Ibid.

[42] Ibid.

[43] Ibid. Koch's fourth postulate—that the organism be found in the tissues, blood, or discharge of the experimentally inoculated animals—was superfluous, Conn thought, in the study of soil bacteria if one used sterilized soils.

active organisms may lose their activity."[44] To avoid these mistakes it would be necessary to corroborate the cultural methods with other types of methods—possibly using the microscope, or entirely new methods.

Conn developed and made public his ideas in the context of an extended debate with Selman Waksman on the "relative importance of fungi and bacteria in the soil."[45] Here Conn criticized Waksman's conclusion that fungi live naturally in the soil and produce mycelia there. While Conn agreed that fungi might live for short period in the soil, he thought the proper question to ask was "whether the mycelia were abundant enough in the soil to compare in its activity with the soil bacteria."[46] Conn was especially critical of Waksman's method, which was a simple test of the fungi's ability to grow in soil clumps on agar plates. In his own research Conn avoided what he called "Waksman's indirect method," and instead relied on a combination of microscopic observation and staining techniques.

It took Conn years to discover a combination of techniques to replace the methods exemplified by Waksman's approach. In order to produce direct evidence—and thus avoid Waksman's presuming conclusions—Conn relied on microscopic observation. To enhance this technology for studying microorganisms, he spent 2 years testing the capabilities of various dyes for microscopic analysis. He succeeded in locating a dye that would stain the microorganisms but not the particles of dead organic material that always accompanied them in the soil. Although Conn developed his method to prove Waksman wrong, he believed it had broader significance for soil bacteriology. Viewing the problem from a more general perspective than his laboratory investigations, Conn questioned the current status of knowledge in all of microbiology. Ultimately, he wondered whether anyone had ever "definitely proved that any particular microorganisms [caused] any of the well known biological activities of the soil."[47]

Vinogradskii encountered Conn's method while developing his own direct method. He shared Conn's desire to set soil microbiology on a scientific foundation based on new rigorous methods, and incorporated Conn's dying techniques into the direct method. He subsequently developed it into an extensive and flexible approach for achieving the goals Conn had laid out, ultimately by moving away from a Kochian method. The techniques and terminology Vinogradskii assembled into his "direct method" became a format for his next series of investigations. He used this method to attack the questions he had identified in 1919 as the primary problems in agricultural microbiology. Focusing his efforts on increasingly specific questions, he adjusted his method and language to address the challenges of this research. His answer to this challenge led him to translate the cycle of life concept into an ecological perspective.

---

[44] Ibid., 254

[45] H. Joel Conn, "The Relative Importance of Fungi and Bacteria in the Soil," *Science*, N.S., Vol. 44, Issue 1146 (December, 15, 1916), 857–858.

[46] Conn, "The Relative Importance of Fungi and Bacteria in the Soil," 857.

[47] H. Joel Conn, "The Proof of Microbiological Agency in the Chemical Transformation of Soil," *Science*, N.S., Vol. 44, No. 1185, 252–255. See 252 for quote.

Over the next 30 years, Vinogradskii published a series of variations on the direct method. These variations, which appeared in reports, speeches, and his final monograph—represent stages of his development as an ecologist. During this time, he explored numerous ideas and described them using a wide variety of terms. His lexicon evolved from one founded in the language of biology, chemistry, and physics to one that simultaneously drew upon and redefined the terminology of ecology. Three key concepts are traceable from his 1923 explication of the direct method to his 1949 rewriting of his earlier contributions: First, he continued to study the biological aspects of the soil. Second, to study the biology of the soil—the microflora or "florule" as he called it—he applied the complex of chemical and physical techniques that made up his direct method. Third, he increasingly moved away from trying to isolate the microflora (the biological component of the soil) from the soil's physical and chemical characteristics (its particles and processes, respectively). As he fine-tuned his experiments to investigate microbial nutrition more precisely, he developed the direct method into a synthetic approach that could account for nature's biological, chemical, and physical characteristics in a single system.[48] By the 1940s, he had honed the less-than-perfect direct method into a comprehensive ecological method.

The Direct Method in 1925: The Rise of Soil Microbiology

In 1925, Vinogradskii began to identify soil microbiology as a discipline distinct from general microbiology. His 2-year investigation testing the direct method had bolstered his faith in it, encouraging him to promote it as the ideal approach for transforming soil microbiology into a rigorous field of study. He defined this new discipline as "the study of soil microbes and their activity *in the midst of their environment (au sein de ce milieu)*, their natural environment."[49] In a critical review of the problems and methods in soil science, he identified two clear lines of research in the recent history of soil science. One was microbiology "*senso stricto*"—the study of microbial species that had been isolated from the soil—and the other "biochemistry of the soil"—that is, the study of the chemical processes that take place in the soil itself, under the influence of the microbes.[50] Neither line satisfied him, however, and he adopted a new approach.

[48] For example, Jadwiga Ziemiecka and Lars Romell worked with Vinogradskii in late 1920s, and no doubt brought their own perspectives to the laboratory and its conclusions.

[49] Serge Winogradsky, "Études sur la Microbiologie du Sol, 1. Sur la Méthode," *Annales de L'Institut Pasteur*, Vol. 39, No 4. April 1925, 299–354; for quote see 299.

[50] Ibid., 300. S. I. Kuznetsov later extrapolated Vinogradskii's division between the study of microbes themselves and the study of their role in the soil into ecological language that reflected two aspects of mid-twentieth century ecological microbiology. He portrayed Vinogradskii as the founder of this ecological direction in microbiology. See S. I. Kuznetsov, *Razvitie idei S. N. Vinogradskogo v oblasti ekologicheskoi mikrobiologii* (Moskva: Nauka, 1974), esp. 3–6.

Here he grappled with a typical problem for biological experimentalists—bridging the gap between the laboratory and nature. For Vinogradskii, soil microbiology lacked a comprehensive method that combined laboratory experiments with the wildness of nature. Microbiologists, for example, had studied microbes in determined and extremely varied conditions of artificial cultures; however, they had followed microbial activity in their natural environment "only mentally."[51] Soil biochemistry, on the other hand, was too limited in scope because it studied only the ultimate chemical effects that occurred in the soil without ever considering the agents indicated by microbiological investigations (the microflora). Consequently, even the combination of these two approaches (he thought of them as categories of facts) could produce no substantial results because they did not rely on direct experiments. They could offer, thus, only weak hypothetical conclusions.[52]

For Vinogradskii, these methods were not sufficiently biological. He thought that they might be useful, however, in a "mutually complementary" arrangement, that is, if he could unite them around a technique that provided direct access to biological phenomena. In the 1890s, he had developed such a method in the form of his elective cultures. All methods, however, evolve according to changing research agendas. Even the elective culture method that he and Martianus Beijerinck had used effectively to discover the nitrifying microbes now seemed to him fatally flawed.[53] With a new synthetic method, he hoped to identify "grand general phenomena" in the chaotic noise of microbial life in the soil. These phenomena represented, for him, "the central problems in soil microbiology" that he had demarcated in his 1920 literature survey: nitrification, nitrogen fixation, ammoniafication, the decomposition of organic matter, and eremacausis (the slow combustion of humus-formed matter).[54] With this goal in mind, he combined the elective cultures with the sophisticated chemical analyses of soil biochemistry, and he breathed new life into his hoary technique.[55]

At the heart of Vinogradskii's criticism of the elective culture method lay his new respect for the process of adaptation or, as he called it, the "selection process."[56] He had supposed that after cultures had run through the course of their life cycle to "biological reaction," they should enter a state of rest. In some, however, he (and others) had observed a rapid increase in microbial proliferation despite the maintenance of the original cultural conditions. This led him to conclude that "adaptation"

[51] Serge Winogradsky, "Études sur la Microbiologie du Sol, 1. Sur la Méthode," 1925, 300.

[52] Ibid., 299–301.

[53] Ibid., 307. Vinogradskii believed that Beijerinck's "accumulation experiments" were exactly the elective culture methods he had developed called by another name.

[54] Serge Winogradsky, "Études sur la Microbiologie du Sol, 1. Sur la Méthode," 1925, 301–306. Liebig had coined the term eremacausis in the 1840s.

[55] Ibid.

[56] Although some American bacteriologists had come to call this process of creating new breeds of bacteria "Darwinizing," Vinogradskii purposely avoided Darwinian language of a struggle for existence—he based his notion of competition on a struggle for food. This was very much in the tradition of Russian evolutionary theory.

had produced a new race that could live in the now liquid environment of the culture. In other words, he believed that the incubation process of the elective culture method was in reality a *selection process* that produced an aerotaxic and mobile colony better adapted to an aquatic existence."[57] He was not ready to abandon the elective culture methods, however, because they still provided him an approximate understanding of the original species' properties. They could be used if one carefully matched the cultural conditions with those of the natural environment; this facilitated a reprieve from the struggle for food and excluded any chance of selection.[58] One simply had to be careful to distinguish the *plante de culture* that the laboratory race had become from its wild ancestor.[59]

The flexibility of the soil's biological component led Vinogradskii to expand his definition of "normal conditions."[60] In his review of the field, he had found numerous studies that focused on "the numeration [the statistical estimation] of germs" using plate counting methods. These counting exercises failed, he argued, because they used artificial gelatin-based environments in their cultures. They could not provide an accurate count of all microbes living in the soil, but rather only of the few species that could grow in gelatin. Moreover, in the wild conditions found in *normal soil*, this small population survived most often in its inactive form. To obtain the actual and total quantity of soil microbes—that is, a closer understanding of the microflora—he proposed using a "direct microscopic study [that] gives us a visual image of the soil population."[61] For him, the profoundly varied functions of different microbial species living in the normal soil made it impossible to count them solely by growing them in standard environments. That is, using these pure culture methods left soil microbiology quantitatively imprecise. There was, furthermore, a qualitative drawback to this approach—although a large number of species had been isolated with such conventional methods, these were limited to species well adapted to the standard environments.[62]

In the reports that appeared during his early Brie-Comte-Robert years, Vinogradskii discussed a wide number of topics and published a substantial amount of new data on the chemical nature of soil microorganisms. I have focused on the aspects most essential for understanding how he transformed his direct method from a biological to an ecological approach. In his 1925 report "Études sur la Microbiologie du Sol, 1. Sur la Méthode," he divided his direct method—his global approach that I associate with his cycle of life perspective—into three groups of techniques. As we will see, he organized this complex of methods around the concept of the microflora; that is, around the biological entities he had been investigating since his apprenticeship with Famintsyn.

[57] Ibid., 307–308.

[58] Ibid., 309.

[59] Ibid., 308–309. The new race either had acquired new properties of the culture or had lost certain characteristics of the original species.

[60] Ibid., 309.

[61] Ibid.

[62] Ibid., 309. Most often these species were sporogenic bacilli.

Vinogradskii described his preferred methods for investigating the micro-"biological" nature of the soil. His first technique—the microbiological microscopy of the soil (*terre*)—would improve the study of the microflora by making possible "a direct microscopic examination as indispensable and sensitive as those already in use."[63] His second technique—cultures grown using natural soils—reflected his criticism that pure cultures made with conventional gelatin environments gave an "unreliable, imprecise, and hypothetical notion" of microbial species' function in the soil environment.[64] He thought it necessary to substitute artificial cultures with ones "in the state of nature," which then could be enriched with diverse substances according to the function being studied. His third technique—auxiliary cultures in a solid environment that imitates the soil—served his efforts to introduce an environmental formula that reproduced as exactly as possible "the conditions of special[ized] life of the species groups being studied."[65] Outlining his report around these techniques, he also provided some preliminary, yet "important" results. These findings, he thought, played two roles in his work—on the one hand, they resulted from his new method and on the other, they directed his research.[66]

Vinogradskii applied these techniques to study the microflora in its natural environment from three directions: chemical, physical, and biological. At this stage, he separated these as distinctly as possible in his investigation and perhaps even in his mind. Microbiological microscopy allowed him to obtain evidence about the physical environment the microflora inhabited. Using "physical microanalysis," as he called it, he separated or disassociated the elements of the soil.[67] The goal was to separate the microflora from the miniscule particles of minerals and organic material.

Vinogradskii met with mixed success using Conn's method. Conn had prepared his samples by shaking a gram of soil with a mixture of water and gelatin, which produced a mixture sticky enough to be spread in a 1-cm^2 area on a specimen plate. After drying and dyeing it (he had found that the dye 'rose bengal' worked best), he counted the bacteria using a microscope—the number of bacteria in that soil sample could then be estimated. What Vinogradskii liked about this method was that it left the bacteria a dark rose color, the minerals colorless, and the dead organic material light red. However, he found the coloring effect to be inconsistent, often leaving the organic material yellow or colorless—that is, they were the color of the bacteria.

The method Vinogradskii offered as an alternative to Conn's did more than reduce the chance of adding too much gelatin or leaving the dye on too long. He developed a complex procedure that incorporated the use of a centrifuge to separate physically

---

[63] Ibid., 315–316. There is an interesting translation issue with the work *terre*, which means either earth or soil. As we will see Vinogradskii also used *sol* (dirt). Perhaps the word *terre* translated as earth is more appropriate when treating his search for a global, general definition of the microflora.

[64] Ibid.

[65] Ibid., 316.

[66] The full report of his findings appeared in the *Annales* in 1926.

[67] Ibid., 319.

the microflora from the crowd of soil minerals and organic matter that surrounded it.[68] Using this method, he peeled away layers from nature's complexity, baring a series of microbial environments for his scrutiny. Even before spinning the soil suspension—a dilution of water and soil—a succession of sediments formed beginning with the coarser materials and growing increasingly fine. Finally, he used the centrifuge—at varying intensities and durations—to "microanalyze" the soil, that is, to separate the soil elements according to their mass and weight—producing very clear preparations containing less and less debris.

He viewed this process not only as a way for microbiologists to obtain clear suspensions of microbes, but also as way to "get an idea of the distribution of the microbes in the soil."[69] This knowledge, he thought, would help them know whether microbes grew on large or small mineral particles, organic or mineral colloids, or if they "simply grew in the disintegrated soil like in a culture broth (*bouillon de culture*)."[70] In this description of the interaction between the microflora and its physical environment Vinogradskii demonstrated his consistent vision of the importance for considering natural places in laboratory investigations.[71]

Microscopic analysis of the suspensions produced by centrifuging showed Vinogradskii that very few microbes lived on the various miniscule particles. The environment that remained, therefore, was the disintegrated soil (or soil "dissolution," as he called it). To investigate this environment, he ran a centrifuged suspension that contained only very fine particles and "the majority of microbes that were able to grow there and that had not been carried away by the centrifugal action."[72] These suspensions provided the materials for further microscopic analysis. It was here that he again introduced the concept of a control soil.[73]

Only the most perfect microscopic method, Vinogradskii thought, could serve as the basis for a successful experimental method. Success was determined by "establishing a stable microbial population—a population type—which would serve as a point of comparison for other species and soil environments.[74] First, he needed to find a "normal soil" and study its microflora carefully.[75] Then, using this control soil he could monitor changes in the microflora as he altered its environment. Studying the characteristics of this microflora allowed one to "lay out a clearly defined biological

---

[68] Ibid., 319–320. As hard as he tried, he could not prepare a suspension of the complete soil sample in all its messy wild complexity—it was never possible to see the microbes as clearly as one could using the pure cultures.

[69] Ibid., 322.

[70] Ibid., 323.

[71] At a point in the early 1930s, this tension forced Vinogradskii to synthesize the language of biology, chemistry, and physics (or physical geography) and to express his ideas as ecology.

[72] Ibid., 328–329.

[73] Ibid., 330

[74] Ibid., 329.

[75] Ibid.

state of the soil, which could be modified experimentally."[76] As he had outlined in his 1923 direct method, he would then provoke a "biological reaction," producing new growths or cultures in the control soil.[77] He "superimposed" these new growths on the "stable florule" thus modifying the microbial population both quantitatively and qualitatively. He measured this reaction by direct microscopic observation. Although this approach seems identical to his 1923 version, attempting to interpret his observations in this new biological framework, he came to rely more heavily on the biological reaction. This reaction "imposed itself on the observer," he wrote, "because the new forms appear very different from the original florule."[78] He had impressed himself with both the great amount of variation he could extract from wild nature and his ability to study it systematically.

Vinogradskii, then, brought the wildness of nature into his laboratory in the form of normal soil. Only what was normal in nature could be a control in the laboratory. But where to find a normal soil? He had not far to look; the land of his own Brie-Comte-Robert estate met all his criteria of normalcy.[79] On his 40 acres, he found a soil that had not been fertilized with manure or plowed for at least 3 years. In order to elaborate a general method for soil microbiology, however, he needed more that a single sample of normal soil. It was important to find other soils from diverse provenances that "shared the same characteristics and stability of microbial flora" as the Brie-Comte-Robert sample.[80]

Elaborating a general method meant making his observations of local microflora meaningful on a global scale. To accomplish this, he used the approach he had developed in the nitrification investigations that led to his discovery of chemosynthesis. Drawing on his extensive network of colleagues—extending to Selman Waksman in the United States, G. Truffaut in Tunisia, and Lebediantsev in Russia—he collected a diverse set of soil samples from around the world.[81] He was not interested in a detailed analysis of these samples. Most important, rather, was that upon microscopic inspection none possessed a microflora with any peculiar characteristics. Across different origins and soil types, from a microbiological perspective the microflora was an undifferentiated feature.[82]

Vinogradskii did not isolate these microfloras, however, in their physical "ambience."[83] Of the twelve samples he examined completely, he selected five for

---

[76] Ibid., 330.

[77] Ibid.

[78] Ibid. Here Vinogradskii, in effect, turned the microflora into the testable part of the soil—he now classified soils as either active or inactive according to the action of the microbial population.

[79] Ibid., 331

[80] Ibid.

[81] Ibid., 331–332. He received a total of 28 samples of French soil from Alfred Bruno of the French Institute of Agronomic Research, twelve North American soils from Selman Waksman, five Tunisian soils from G. Truffaut, and five Russian soils from Lebediantsev.

[82] Ibid., 333.

[83] Ibid., 333.

special treatment in his discussion—those from Somme, Bas-Rhine, America, Russia, and of course his perfect soil from Brie-Comte-Robert. Referring his readers to the color plates (drawings) that accompanied this publication, he highlighted the "microscopic picture of the florules of these five soils with all their ambience drawn according to nature."[84] Describing his drawings—in which he tried to reproduce reality as close as possible—he stressed the disposition of the colonies, which were very close together and very numerous.[85] He defined the ambience he observed in his suspensions as a microbial habitat. In his opinion, this physical habitat should be of primary importance to any scientist studying the soil—not only to mineralogists and pedologists but also to microbiologists.[86] He claimed that his method would help study "the ultimate elements of the soil outside of the microbes that live there."[87]

Vinogradskii's microscopic study of the control soils demonstrated the existence of two large groups of soil organisms, which differed by their general functional characteristics. The first group, the zymogenes, only propagated in soils that contained the essential fermentable substances related to the process being studied. In addition, their presence in the soil caused these processes to run at an intense rate. On the other hand, the second group grew normally in soil poor in these substances and "function as an agent of slow combustion of humus material."[88] From a functional point of view he called these bacteria humivores, and from an ecological perspective he called them autochtones, or aboriginal species.

Conn's concept of "watchful waiting" resonated with Vinogradskii's plan to study the relationship between these two groups.[89] Vinogradskii thought that by studying the activity of these agents of slow combustion, it would be possible to determine the cause of all the different processes.[90] This conception of nature divided

---

[84] Ibid. He drew these on a scale of 1 μm to 2 mm, or magnified 2,000 times.

[85] When making this point about the number of colonies, he pointed out that four times more colonies appeared than when viewing the same soil samples magnified only at 1,000 diameters. A reading of Vinogradskii's drawings reveals his careful placing of the microbial communities in spaces separated by the physical particles that formed natural barriers. Someone not familiar with this microscopic view of the soil may even find the careful arrangement of the "wildness" humorous. [An interview with his niece in 2006 leads me to believe that he was not known for his sense of humor.]

[86] Ibid., 333–334.

[87] Ibid., 334.

[88] Serge Winogradsky, "Études sur la Microbiologie du Sol, 1. Sur la Méthode," 341. Here he referred his readers to his 1924 article in the *Comptes Rendus*, No. 178, (7 April 1924).

[89] Ibid., 341.

[90] He did note that even though future studies might discover organisms of ambiguous function that were transient between the two groups, this would not obliterate his delineation of these two groups. Vinogradskii pointed out how the two remaining parts of the direct method would facilitate this investigation of the microflora biologically: (1) cultures in the natural soil and (2) and auxiliary cultures. Cultures made with the natural soil required no environmental preparations or inoculations—to fresh, control soils with all their germs were added energetic substances. These "broke" (*rompre*) the relative biological equilibrium and produced the growth of the latent zymogenes.

into fast and slow agents of combustion substantiated again his belief that "studying such specialized functions demands methods just as specialized and adapted to these functions."[91] At this time, Vinogradskii still separated physiological processes from their ecological function. When these two merged in his imagination and he began to develop a global ecological perspective, microbial roles in nature became central to ecology, and ecology came to represent the cycle of life.

[91] Ibid., 341.

Chapter 8
Ecological Microbiology

Vinogradskii first addressed the subject of ecology in his discussion of soil microflora. Attempting to alleviate the haphazard nature of soil microflora research, he proposed an ecological definition of that field. Envisioning the soil as "a dumping ground for all waste," he assumed that all germs would be able to live in it at some time and in some place depending on their nutritional needs and the local soil conditions. The category "soil microflora" would thus necessarily include almost all known species, ranging from the nitrifiers, fixers, humivores, and other autochthonous microbes to the bacteria buried with cadavers. In order to make ecological sense of this wide variation and local dependency of the soil microbial population, he proposed limiting the term "soil microflora" to encompass "the microbes especially adapted to a life in the soil environment."[1] Recognizing that this environment could be—and in his experience often was—"dirtied" (*souille*) by foreign decomposing materials, he clarified that by "soil" he meant "biologically normal mineral soil."[2] Although this type of soil does not contain organic substances in the process of fermentation, some substances may have begun to degrade to a comparatively stable stage. At this stage of development, the soil contains substances known as *humiques*. It was only here, in this normal soil that the true microflora lived.[3]

Vinogradskii pointed out that the currently incomplete understanding of the microflora could be remedied only by creating a method to analyze the soil microbiologically. In comparison to the rather simple understanding of this problem in the early days of microbiology, its complexity had begun to reveal itself. The "variety of the soil microbes' functions," he thought, made current methods impotent and too conventional—they had been "reduced to several formulas that were insufficient for studying natural phenomena."[4]

---

[1] Ibid., 310. Here he called his readers' attention to his "Sur la Microflore Autochtone de la Terre Arable," *Comptes Rendus*, Vol. 178, April 7, 1924.

[2] Ibid.

[3] Ibid.

[4] Ibid.

L. Ackert, *Sergei Vinogradskii and the Cycle of Life: From the Thermodynamics of Life to Ecological Microbiology, 1850-1950*, Archimedes 34,
DOI 10.1007/978-94-007-5198-9_8, © Springer Science+Business Media Dordrecht 2013

Vinogradskii was not the first to criticize current microbiological methods for studying the soil. In his survey of the literature, he had discovered that in 1909 Hugo Fischer wondered whether an effective method existed for studying soil bacteria.[5] In addition, Sir John Russell, director of the Agriculture Experiment Station at Rothamsted, had noted that scientists knew very little about the "micro-organic population of the soil."[6] Vinogradskii was especially interested in Russell's comments regarding the inappropriate use of culture methods. Russell had pointed out that new members of the highly complex soil microbial population are "picked out by some culture method, and their physiological effect is studied in an arbitrary culture solution."[7] He thought that this culture method was defective—especially if its results were applied directly to managing the soil—because "microorganisms are considerably influenced by the medium in which they happen to find themselves," and may behave in one way under one set of circumstances but quite differently under other circumstances.[8] Where Russell interpreted this as the organism's ability to respond to different environments, Vinogradskii understood it as adaptation or the "variability of species."[9] Like Vinogradskii, Russell believed that "most micro-organisms exist in two states: an active or trophic state, and a resting state."[10] Because Russell was interested more in how micro-organic populations in the soil influenced plant growth, however, he thought it "reasonable to suppose that the resting forms are comparatively unimportant."[11] They resting forms were not to be regarded as part of this micro-organic population.[12] Vinogradskii disagreed. For him, the resting forms signified the "possibility of latent life, of an inactive stage of these species in the soil."[13]

Vinogradskii attributed the independent development of these ideas to Conn. In 1917, Conn published an article "The Proof of Microbial Agency in the Chemical Transformation of Soil," in which he questioned the validity of common assumptions about the bacteriology of the soil. Had it, he asked, "been definitely proved that any particular microorganisms cause any of the well-known biological activities in soil?"[14]

---

[5] Serge Winogradsky, "Études sur la Microbiologie du Sol, 1. Sur la Méthode," 1925, 311. Hugo Fischer "Besitzen wir eine brauchbare Methode der bakteriologischen Bodenuntersuchung," *Zentralblatt für Bakteriologie, Parasitenkunde und Infektionskrankheiten*, 1909, Abt. II, No. 23, 144–159; see 144.

[6] Serge Winogradsky, "Études sur la Microbiologie du Sol, 1. Sur la Méthode," 1925, 311. See Edward J. Russell, *Soil Conditions and Plant Growth*, *The Rothamsted Monographs on Agricultural Science*, Dr. E. J. Russell, ed., (London: Longmans, Green and Co., 1921), 4th Edition, see esp. 250–251.

[7] Russell, *Soil Conditions and Plant Growth*, 250.

[8] Ibid.

[9] Serge Winogradsky, "Études sur la Microbiologie du Sol, 1. Sur la Méthode," 1925, 311

[10] Edward J. Russell, *Soil Conditions and Plant Growth*, 250.

[11] Ibid.

[12] Ibid.

[13] Serge Winogradsky, "Études sur la Microbiologie du Sol, 1. Sur la Méthode," 1925, 311

[14] H. Joel Conn, "The Proof of Microbial Agency in the Chemical Transformation of Soil," *Science*, N.S., Vol. 46, Issue 1185, (14 September 1917), 252–255; see 252 for quote.

Conn was troubled by the frequent occurrence of "loose statements" concerning the relationship between certain bacteria and certain chemical transformations in soil, although the "causal relationship had never been obtained."[15] He understood that "it was practically impossible to obtain direct evidence as to what actually goes on in soil." Laboratory tests carried out on pure cultures show only what happens in the laboratory and not in the soil. For Conn, "the activities of bacteria in soil are associative actions; and an organism capable of vigorous activity in pure culture may be almost inactive in the presence of its natural rivals."[16]

Ecological Microbiology in 1925

Vinogradskii's investigatory direction at Brie-Comte-Robert resulted from his desire to rekindle his scientific career. He had outlined new goals and themes for his research in the notes he took during his literature survey of 1920–1921. Although his new projects were founded on his earlier research, he had been shown how to think of that work differently, in the lexicon of a new science—ecology.

He had, for example, become intrigued especially by one article in the *Zentralblatt fur Bakteriologie* in which the author, A. Krainskii, a Russian working in the Microbiological Laboratory of the Technical College in Delft, portrayed Vinogradskii's research and methods as ecology. Moreover, Krainskii placed Vinogradskii in a lineage of ecological investigators that included Pasteur and Beijerinck (Krainskii's mentor at the time). In "The Actinomycetes and their Significance in Nature," Krainskii drew on S. Tschulok's *System of Biology* (1910), in which biology was divided into seven disciplines.[17] It was the fourth, ecology that caught Vinogradskii's attention.[18] Finding Krainskii's work "excellent, clear, thorough, and an excellent exposition," in his notes of this work Vinogradskii copied only the author's definition of ecology—"the adaptation (*Anpassung*) of organisms to the external world (*Aussenwelt*)."[19] Krainskii believed that microbial ecology had been studied "with the use of the elective and accumulation methods," and they yielded "not only important diagnostic species characteristics, but also served as a means for maintaining certain species."[20]

---

[15] Ibid.

[16] Ibid., 252–253.

[17] A. Krainskii, "Die Actinomyceten und ihre Bedeutung in der Nature," *Zentralblatt für Bakteriologie, Parasitenkunde und Infektionskrankheiten*, Abt., II, Vol. 41, No. 6, 649–688. He cites S. Tschulok, *Das System der Biologie: Forschung und Lehre* (Jena: G. Fischer, 1910), 197.

[18] Serge Winogradsky, *Bibliography of the Zentralblatt for Microbiology, 1895–1920*, Service des Archives de l'Institut Pasteur, Box. Win 3, unnumbered pages.

[19] Ibid.

[20] A. Krainskii, "Die Actinomyceten und ihre Bedeutung in der Nature," 650. Vinogradskii had developed the elective culture method and Beijerinck the accumulation method. In the 1920s, Vinogradskii criticized Beijerinck for using the elective culture method and calling it the accumulation method—they are basically the same method.

For Krainskii, ecology was the study of the relationship of an organism to other living things and to inorganic nature by "investigating the conditions of the struggle for existence."[21] Ecological studies of higher organisms existed mainly as the recording of the conditions under which an organism develops in nature. It was very difficult, however, to inventory these conditions for microscopic organisms, rendering the standard ecological methods useless in microbiology. Krainskii traced the development of a new ecological method for microbiology to Pasteur's fermentation investigations. Pasteur's "accumulation methods" helped to create laboratory conditions in which the experimental organism "was allowed to be victorious in the struggle for existence."[22] Krainskii recognized that Vinogradskii had applied Pasteur's method in his nitrification investigations as an "elective method," in which a culture is called elective if it promotes the uncovering of a certain function, which is as restricted as possible.[23] In such a culture, the growth of other microbes is either impossible or is possible only with difficulty. The investigator is, therefore, "helping the sought after microorganism in its competition for life (*Lebenskonkurrenz*) with other microorganisms," the result of which is its accumulation.[24] It is from his reading of Krainskii, as tenuous the definitions or links might have been, that Vinogradskii first came to see himself as an ecologist.

Vinogradskii exercised his new ecological imagination during an exchange of letters with Russell between 1924 and 1938. Here he first described his approach in ecological terms. After the international conference in Rome (1924), Russell sent Vinogradskii a French translation of his *Soil Conditions and Plant Growth*, a book Vinogradskii had cited in his recent publications. Russell's admiration for Vinogradskii is clear in this first letter—he requested that Vinogradskii visit the Rothamsted laboratories and asked for a signed photograph. Moreover, Russell thought that Vinogradskii might offer suggestions for his research group's current "soil microbiology research" and he commented that "we find your new method of exploring the soil micro-organisms very useful and I am hoping that much will be learnt as a result of applying it."[25]

Russell's admiration for Vinogradskii did not prevent him from defending his group's approach. In November 1927, Russell responded to Vinogradskii's criticisms of his use of pure cultures, his assumption that the biochemical processes carried out by the soil microorganisms in laboratory media will also take place in the soil, and of his use of plating as a means estimating numbers of bacteria in the soil.[26]

[21] Ibid., 656.

[22] Ibid. In his experiments, Krainskii thought, Pasteur had applied the accumulation method by "infecting" a sterile liquid with a fermenting liquid at the peak of its fermentation activity.

[23] Ibid., 656–657.

[24] Ibid., 657.

[25] Russell to Vinogradskii, October 21, 1924. Archives de l'Institut Pasteur. Box Win 2, Folder Correspondence International, Angleterre, Russell. On Russell see Richard P. Aulie, "Edward John Russell," *Dictionary of Scientific Biography*, Vol. 12, 492–493.

[26] Russell to Vinogradskii, November, 28 1927, Archives de l'Institut Pasteur. Box Win 2, Folder Correspondence International, Angleterre, Russell.

Russell had always agreed that pure cultures represented the best method of studying the soil microorganisms, and especially for uncovering their physiological possibilities and their reactions to various external conditions. He admitted however "that microorganisms have a fairly wide range of adaptation, and that in the soil, where conditions are entirely different from those of culture media, the reactions effected by the microorganisms may differ also."[27] Because many species of microorganisms were "either of undistinguished morphology or could alter their morphology considerably under changing conditions," they could be identified only after isolating them in pure culture.[28] He described the current method in use in his laboratory—first, to isolate as many organisms as possible that will attack the compounds studied in pure culture; second, to ascertain whether the organisms will operate in sterile soil, and third, to add the compound to field soil and see which of the organisms is caused to increase. To measure this last phenomenon, his group used the plating method Vinogradskii had found so offensive.

Vinogradskii's criticism of the plating method surprised Russell since he and his group had received good results with it. Although they had used the plating method to estimate bacterial numbers, they agreed with Vinogradskii that these estimations in no way represented the total number of bacteria in their soil. They used the plating method only to measure relative changes in bacterial numbers, not to provide absolute numbers. Agreeing that many bacterial forms did not appear in the plating method, Russell argued that "it seemed likely that anything that would raise or lower the numbers of bacteria in the soil would raise or lower the numbers appearing on the plate."[29] Although Russell stood by his plating method for estimating fluctuations in bacterial number, he was "keeping in close touch with the developments in the way of direct microscopic examination of the soil" and was studying it "from the statistical point of view in hopes of learning more about the distribution of soil microorganisms."[30]

In his response to Russell's defense of the plating method, Vinogradskii reiterated the strong distinctions he drew between general microbiology and soil microbiology. Not only did he agree that pure cultures were the best method for studying "the physiological *possibilities* of the organisms and their reactions to external conditions," he considered it the only method.[31] When studying their "*actualities*" in a natural medium—the real aim of soil microbiology; however, this artificial method ceased to be reliable. They should instead use any other method that monitoring "the free play of microbial activities under conditions as close as possible to the natural ones."[32] Reacting to Russell's contention that "a separate study of soil microbiology

[27] Ibid., 1–2.

[28] Ibid., 3.

[29] Ibid.

[30] Ibid.

[31] Vinogradskii to Russell, (undated, but most likely written soon after November 28, 1927). Archives de l'Institut Pasteur. Box Win 2, Folder Correspondence International, Angleterre, Russell, 1.

[32] Ibid.

is a mere matter of convenience," Vinogradskii argued that there was "a fundamental distinction between general microbiology—the study of microscopic morphology and physiology—and soil microbiology, which is properly speaking a microscopic ecology."[33] The "elaborate method" Russell used to complete and control his pure culture data, might, Vinogradskii admitted, occasionally provide good results. Yet, it was too complicated to serve as a general method. For example, the daily bacterial fluctuations Russell had discovered at Rothamsted, while interesting, were difficult to understand and applicable only to a small group of the soil population. Russell's data revealed nothing about how the "unknown conditions" affected the density of the whole population.[34]

The Direct Method in the Late 1920s

Now an ecologist at heart, Vinogradskii made his new direction clear in his published work. Applying his direct method to investigate a practical agricultural question, he sought to determine the causes and characteristics of nitrogen fixation in the soil.[35] Working with a new visiting assistant, Jadwiga Ziemiecka, he expanded his investigation of the nitrogen fixing power of soils.[36] Nitrogen fixation remained an important and popular topic in microbiology since its reformulation as a biological (rather than chemical) phenomenon in the 1890s.[37] Vinogradskii set out his new direction, correlating it to his use of new methods. Since he had executed his investigation "using new methods," it could not "be considered as an immediate continuation of the previous line of researches."[38] In this new direction, he strictly maintained his ecological focus on "the microbiology of the natural environment" and ignored what other scientists had learned about nitrogen fixation using pure culture methods.[39] In some ways his ecology was similar to plant physiology—in both he investigated the role of the microbes in biochemical soil processes. The language of ecology, however, freed him to equate the microflora with those processes, and, conversely, to characterize the soil solely in terms of its microflora.

[33] Ibid.

[34] Ibid., 2.

[35] Serge Winogradsky, "Études sur la Microbiologie du Sol: Sur les Microbes Fixateurs d'Azote," *Annales de L'Institut Pasteur*, Vol. 40, No 6. Juin 1926, 455–520. In 1926 in his second article on soil microbiology, Vinogradskii reported on his investigations of nitrogen fixation.

[36] Serge Winogradsky en collaboration avec J. Ziemiecka, "Études sur la Microbiologie du Sol: Sur le pouvoir fixateur des Terres," *Annales de L'Institute Pasteur*, Vol. 42., No. 1, 1928, 36–62.

[37] In his report detailing this research, Vinogradskii deferred his readers interested in a historical survey of the question to the list recently complied by his student V. I. Omelianskii. Omelianskii had published a list of 430 publications in his 1923 literature review on this topic.

[38] Serge Winogradsky, "Études sur la Microbiologie du Sol: Sur les Microbes Fixateurs d'Azote," 455.

[39] Ibid., 456.

The institutional and intellectual demands of Vinogradskii's nitrogen fixation investigations led him to stress the flexibility of his direct method. In response, he expanded and refined one technique that had played a relatively small part in his direct method of 1923–1925. Trying to find a cheap, fast, and effective method for agriculturalists to test their soils for nitrogen fixing ability, he concentrated his efforts on silica gel plates. He altered his direct method to address this practical question, maintaining, however, his goal of discovering the general phenomena that governed the microbial world.

Nitrogen fixation research had attracted the attention of some of the founders of microbiology. Pasteur, and Trecul and Van Tiegham had isolated organisms that were able to fix atmospheric nitrogen in the soil. Because these organisms were able to grow in nitrogen-poor and hydrocarbon-rich environments, they were eventually classified as a physiological type—nitrogen fixers.[40] Vinogradskii focused his research on two of these—the anaerobic microbe *Clostridium Pastorianum*, and the aerobic microbe *Azotobacter* discovered by Beijerinck. Their fixing actions had been determined in laboratory investigations, yet the observation most significant to Vinogradskii was that the germs of these organisms appeared very widespread in the soil.[41] He criticized, however, other microbiologists who attributed their laboratory findings to the organisms living in the natural environment without conducting a special study of "the conditions that preside over the activities of the microbes, conditions that a flask filled with liquid is too far from reproducing."[42] These scientists falsely assumed, he argued, that a soil's ability to exhibit butyric fermentation revealed nothing about the density of microbes in that soil. Moreover, this test ignored another crucial question—whether the microbes could move from their spore to active state—that is, the "biological reaction" in Vinogradskii's direct method.[43]

Vinogradskii applied his ecological methods to uncover the complex dynamics occurring in nature that caused microbes to fix nitrogen. Although recent studies using the new microscopic methods had shown that spontaneous proliferations of one nitrogen fixer, *Clostridium* did in fact occur in natural soil environments—demonstrating their capacity as fixing agents—Vinogradskii remained unimpressed

[40] The trait of these microbes to grow in particular environments had provided Vinogradskii with the basis for studying "the agents of fixation and their processes," that is, his elective culture method of the 1890s. Ibid.

[41] He had found that one could almost without fail produce a butyric fermentation—the fixing process associated with these microbes—by throwing a couple of grams of soil in a sugar solution. Ibid., 457.

[42] Ibid., 457.

[43] Here Vinogradskii sets himself apart from his colleagues because he viewed the relationship between the evidence from the laboratory and that from nature. For a further discussion of this, see Bruno Latour, "The Costly Ghastly Kitchen," in Andrew Cunningham and Perry Williams, eds., *The Laboratory Revolution In Medicine* (Cambridge: Cambridge University Press, 1992), 295–303; and Ibid., *Pasteurization of France* (Cambridge, MA: Harvard University Press, 1988), Alan Sheridan and John Law, trans.

with the state of knowledge of these microbes. He wanted to know the frequency of these processes in the soil, the nature of the soil that either favored or hindered them, the conditions necessary for it to progress in the natural environment, its relationship to other microbial processes, and its economic significance?[44] The only way to answer these questions was with his new "effective methods"—methods not commonly used in microbiological laboratories.

In 1928, Vinogradskii and Ziemiecka put the effectiveness of the direct method to a test.[45] Vinogradskii believed his methods offered a fast and reliable way to determine the total density and state of activity of the fixation processes in the soil. By correlating the ability of a series of different soils to perform certain tasks with the activity of a specific microbial population in the soil, Vinogradskii and Ziemiecka were pursuing an ecological understanding of the soil. Towards that goal, they applied the direct method—reconfigured as a regimen of silica gel plates and spontaneous cultures ("natural cultures in the soil itself")—to investigate the relative density of *Azotobacter*. In this research, they challenged the standard method, which had been in use since Beijerinck developed it to investigate the presence of microbes in the soil and maintain them in cultures. Although this method suited Beijerinck's goals, it was too rudimentary, Vinogradskii thought, to explore more subtle questions about soil dynamics. "To study the distribution of *Azotobacter* in nature and to determine their role there," for example, a method needed to do more than establish the presence or absence of a microbe.[46] It must measure their absolute number, their density, and their state of activity in the natural environment. Unable to satisfy these criteria, past methods were insufficient for studying "*the activity of a specific population within the soil* or the *fixing ability of the earth*."[47] Vinogradskii proposed using his direct method to succeed where others had failed.

The demands of the *Azotobacter* work—to determine their total density in the soil—led him to adjust his method. Transforming the direct method into a precise tool for measuring density, he used large silica gel cultures as an elective culture and relegated the spontaneous cultures to secondary status.[48] At the heart of this approach, was his assumption that every germ, whether in an active or latent state of activity, proliferates in some "perfectly elective environment." Here, again, Vinogradskii continued to correlate each germ with a particular physiological process in a determined environment. With his method he would attempt to separate the multitude of processes occurring in his gram of soil, by surrounding that soil with a control

---

[44] Ibid., 457–458.

[45] Serge Winogradsky en collaboration avec J. Ziemiecka, "Etudes sur la Microbiologie du Sol: Sur le pouvoir fixateur des Terres," *Annales de L'Institut Pasteur*, Vol. 42., No. 1, 1928, 36–62. At the time of the 1928 publication, the Danish forestry student. Lars Gunnar Romell, came to work in Vinogradskii's laboratory. This experience would prove to be a defining moment in Romell's career, leading him to become one of the staunchest supporters of Vinogradskii's ecological methods.

[46] Ibid., 36.

[47] Ibid., 36–38.

[48] Ibid., 39.

environment. The germs in his soil sample would spread out from the sample onto the silica gel plate saturated with the nutrient they required and best accommodated them. A series of these plates would then allow the experimenters to correlate the density of microbes on the plate—which they called the "natural colonies"—with that of a specific population in the soil.[49] This "direct" measure of density served as a crucial "factor for judging the activity of the soil."[50]

Vinogradskii and Ziemiecka also related density and microbial activity to soil and environmental fertility. On their plates they observed that a maximum of microbial density corresponded to an active stage and a minimum to an inactive stage.[51] The difficulty lay, however, in determining whether a specific population at intermediary stages of density were tending upwards (on a curve) or downwards. Spontaneous cultures would allow for testing which element or factor in the culture limited the activity or fertility of the soil. They first measured the total microbial density in a soil sample with silica gel plates, and used that information to determine their state of activity. If the microbes were inactive, they classified the soil as unfertile and conducted a series of spontaneous cultures according to the 1925 direct method. These would reveal the "limiting factor." In the natural environment, Vinogradskii pointed out, the possible limiting factors and special conditions that might affect the fixing ability (fertility) of a soil were numerous.[52]

In this scheme, Vinogradskii stressed, not the chemical or physical factors involved, but rather the ecological activity of the microflora. In a series of silica gel plates inoculated with increasing amounts of nitrite—a known *Azotobacter* "limiting factor"—Vinogradskii and Ziemiecka observed that the *Azotobacter* colonies diminished in number and density. Simultaneously the number of other bacterial colonies and microscopic fungi increased and the amount of nitrogen fixing that occurred decreased. These observations left them with no doubt about the role of the nitrite—"it released an antagonistic biological factor that hindered or removed the activity of the fixers."[53] In an interpretive move very similar to his interpretation of *Beggiatoa*'s role in wild sulphur springs, Vinogradskii assigned the primary role or agency responsible for the natural phenomenon to a specific microbe. Fixation depended on the amount of nitrogen in the environment—yet to measure this one had to "correlate it with the density and activity of *Azotobacter* in the soil."[54]

Committed to an ecological logic founded on the concept of a purposeful and circulatory nature, Vinogradskii considered this "antagonistic biological factor" merely as a "chronic" feature. Even in the direst situations, the common bacteria would drive the *Azotobacter* only to inactivity or a very low density. Waiting in the soil for an environment favorable for their proliferation, the *Azotobacter* would then

[49] Ibid.

[50] Ibid.

[51] Ibid., 40

[52] Ibid., 42.

[53] Ibid., 45–47

[54] Ibid., 47.

"return to the offensive."[55] This "direct" observation of a struggle for food, or energizing material, returned Vinogradskii to his vision of the global nitrogen regimen in the soil environment. By studying the effect of these biological factors on fixation across diverse soils, one could investigate the grand reserve of total nitrogen.

The relationship of a natural soil process to its biological agent, however, was not static. The phenomenon of fixation, for example, was "not the exclusive privilege of one specific species or group of microbes."[56] A particular species, instead, was made "a natural agent or a ferment of a definite process" not by the "quality of a function" but rather by its "physiological productivity" (*rendement*) and even more "by the adaptation of the being to the surrounding environment (*ambiance*), from a biological, chemical and physical point of view. This adaptation allowed the microbe to "invade the natural environment and exercise, regularly and continuously, its maximum of activity."[57] In the late-1920s, Vinogradskii began to transform his direct method into an ecological approach. This led him to express the relationship between a microbe's physiology and its environment in a new way—using the language of experiment to describe the nature of the soil.

The Direct Method and Microflora in the Early 1930s

At the very beginning of the 1930s, Vinogradskii continued to describe his general vision of nature as soil biology, a vision that recalled his cycle of life perspective. In an address on the biology of the colonial soils to the Association of Colonial Scientific Research, he expressed this vision more explicitly that he had in his more formal published scientific reports.[58] Assigned to discuss the state of soil science in the colonies, yet finding that there was no research in progress at all, he limited his report to "demonstrating the utility (one should rather say the necessity) of biological research" for understanding the phenomena that occur in the soil.[59] The time had passed when the soil was considered a "dead mass, composed of organic and inorganic debris." No longer studying the soil only in terms of its geological origin—that is, its chemical and physical characteristics—soil scientists, Vinogradskii thought, had learned much about the statistics of the soil. This research, however, did not address the dimension Vinogradskii considered most crucial for understanding the nature of the soil—its dynamics and the microscopic beings that monopolized them.[60]

Vinogradskii introduced his audience to the idea that the soil could be considered as "a living environment, as a collective entity that possessed the characteristic

[55] Ibid.

[56] Ibid., 50.

[57] Ibid.

[58] Serge Winogradsky, "La Biologie du Sol," *Le Mans* (Paris : Imprimerie Monnoyer, 1931), 1–4.

[59] Ibid., 1.

[60] Ibid.

functions of a living organism."[61] This entity absorbed oxygen and exhaled carbon dioxide—thus it respired; it also digested by degrading complex organic substances as well as simple chemical bodies. This digestion produced excretions, such as entirely mineralized material useful for plants, or "black material, the humus, which resists the chemical energies of the soil and accumulates" in the earth.[62] All the chemical analyses and syntheses that occurred in a living being, also transpired in the soil due to the proliferation and re-proliferation of microbial cohorts responding to the changing supply of energizing material.[63]

Vinogradskii distinguished the microbial population inhabiting the soil from the "horde of species" found in polluted or putrefying waters. This population existed instead as a "real microbial apparatus" (*veritable appareil microbien*) composed of a multitude of species with varied functions, often strictly specific, which remain at rest or enter into action in accordance with exterior stimulants."[64] The well-ordered activity of this microbial apparatus quickly decomposed the animal materials buried in the soil, converting them into materials useful for the plants, especially carbon dioxide. The microbial apparatus then decomposed the plant materials through a succession of fermentations, and broke these materials down into their constituent bodies, eventually forming a new protein-rich substance called humus. Vinogradskii viewed the products of animal and plant decomposition—carbon dioxide and humus—as "the final links . . . in nature's carbon cycle, which was recaptured anew ... with the assimilation of carbon dioxide by the green plants."[65] Similarly, microbes completed the nitrogen cycle, providing plant across the globe their required nutrients, which would otherwise be left inaccessible in nature's inexhaustible reservoir. Only the specific microbes were able to complete this cycle, assimilating the gaseous nitrogen directly from the atmosphere and transforming it into the ammonia that plants and agriculturalists required.[66]

In 1932, Vinogradskii discussed these cycles in his critique of previous investigations that he thought focused too narrowly on the chemical dynamics of the soil, while ignoring the critical role of microbiological processes.[67] The microbiological method made the chemical method more precise by being able to investigate which organisms "attacked" and decomposed which specific animal and vegetable substances in the soil.[68] By clearing up these "modalities of action" he could "allocate the microbes' roles in nature." When in their "microbial community"—where "all

---

[61] Ibid.

[62] Ibid., 1–2.

[63] Ibid., 2.

[64] Ibid.

[65] Ibid., 3.

[66] Ibid.

[67] Serge Winogradsky, "Études sur la Microbiologie du Sol (cinquième mémoire): Analyse Microbiologique du Sol, Principes d'une Nouvelle Méthode," *Annales de l'Institut Pasteur*, Vol. 48, No. 1, Janvier 1932, 89–133 ; See 94–95.

[68] Ibid., 94.

abilities were likely represented"—these agents exercised their "special aptitudes" or "function of predilection."[69] Envisioning this function as the "natural function of a species in the soil," Vinogradskii argued that it could only be studied in an environment of "free competition" in which the sole determining factor was the energizing substance.[70] Not only did he recommend banishing the pure culture, but he also demanded "holding strictly to the conditions operating in the presence of the entire soil microflora without any exceptions.[71] Based on these ideas, he turned the current methods on their heads: rather than making one isolated species proliferate in an environment of a complex formula, he offered a single energizing substance to a mixture of microbes—the same mixture that exists in the soil."[72] Armed with this method, agrobiologists and microbiologists would be able to develop a comprehensive understanding of the "functioning of the soil microflora."[73]

Vinogradskii now promoted the silica gel plates as an efficient method for investigating the natural functions of the soil microbes.[74] With them, he had opened a window on the collective reactions of the microbial community, which he conceived not as a formless heap (*amas*) of species, but as a "true microbial apparatus" that had taken shape over the span of countless centuries. Seeing it as both a complex community and an apparatus, he portrayed the microflora as being organized according to the principle of the division of labor, which expressed itself in the affinities or special aptitudes of its members.[75] At the time of this publication, Vinogradskii considered this conception a mere hypothesis, yet the experiments he conducted with it made it highly plausible to him.

In a report to the *Annales Agronomique*—an applied science journal—Vinogradskii outlined his ecological method for microbiology. In this forum, he attributed the founding of the history of microbiology to the work of: Pasteur, Schloesing and Müntz, Koch, and himself. He hoped to impress his French audience, no doubt, by tracing his intellectual lineage to Pasteur—who first developed "the present ideas of the role of microbes as the agents of soil and water dynamics."[76] His credentials displayed, he proceeded to describe his new ecological method and its significance for agriculture.

For Vinogradskii, to study a microbe's activity in nature now meant to study its ecology—now he was living the vision inspired by his reading of Krainskii. He would seek to revitalize general microbiology by introducing into it the

[69] Ibid., 95.

[70] Ibid., 95–96.

[71] Ibid., 96.

[72] Ibid.

[73] Ibid.

[74] Ibid., 126.

[75] Ibid.

[76] Serge Winogradsky, La Microbiologie OEcologique: ses principes et son précède," *Annales Agronomique*, 1939/41, 1–23. See 1.

"ecological spirit."[77] He traced the origin of his idea of an ecological microbiology to the direct method he unveiled at the 1925 Congress of the International Association of Soil Science in Rome. Rather than trying to work out a compromise between this method and classical procedures of general microbiology, he opted to differentiate a new branch of microbiology.[78] Founded on the ecological principle, all of the procedures of this new branch were aimed at "establishing by laboratory experiments the conditions of existence and the activities of the microbes within their habitats."[79] There was one obligatory rule—"to avoid as much as possible all that is artificial, or conventional, in the contemporary methods of general microbiology."[80]

[77] Ibid., 9.

[78] Ibid.

[79] Ibid.

[80] Ibid.

Part V
The Impact of Vinogradskii's Work

Chapter 9
Science is Ecological and Ecology is Scientific: The Uptake of Vinogradskii's Direct Methods

Since his discovery of chemosynthesis in 1890, Vinogradskii had enjoyed an international reputation as a leading scientific figure and his second career with the Pasteur Institute only enhanced his reputation.[1] There he pushed soil microbiology in a new direction and helped launch the field of ecological microbiology. The enthusiastic reception of his work in the twentieth century was based in part on the enduring significance of his earlier discoveries in the late nineteenth century, especially his discovery of autotrophism and his conception of global nutrient cycles.

In the 1930s, his reincarnation as a "French ecologist" attracted a new group of ecologically-minded scientists.[2] To his long-established international network, Vinogradskii added soil scientists, soil microbiologists, and forestry and marine scientists. In search of a method for investigating the role of microbes in their natural conditions, they thought they had found one in Vinogradskii's direct method. His approach attracted them because it satisfied their need for an experimental approach to holistic, ecological questions. It is no wonder that Vinogradskii's method served this purpose—he had developed it for the same reasons when ecology was just emerging from plant physiology. Early twentieth century biologists employed Vinogradskii's approach to make their science more "experimental, analytically

[1] Vinogradskii's legacy extends beyond the list of methods, of various *Microbes vinogradskii*, or his many self-proclaimed protégés. A person of diverse talents, he influenced people and events outside of the scientific community including his family, friends, and even perhaps the political thoughts of a limited readership. It was in science, however, that he made his most enduring impression. Like Schrödinger's cat, the exact nature of his influence changes as new soil scientists, microbiologists, and historians continue to examine and reinterpret his work.

[2] Vinogradskii's reputation in the 1930s is intertwined with previously established fame for the discovery of autotrophism. His earlier work still captured the imagination of twentieth century plant physiologists and microbiologists.

L. Ackert, *Sergei Vinogradskii and the Cycle of Life: From the Thermodynamics of Life to Ecological Microbiology, 1850-1950*, Archimedes 34,
DOI 10.1007/978-94-007-5198-9_9,

rigorous, and integrative."[3] Above, we found that ecology did not become a discipline so much as disciplines became ecological. Vinogradskii's laboratory-based, experimental methods—built as they were upon the cycle of life perspective—thrived in this new environment.

To treat comprehensively how Vinogradskii's methods migrated through the scientific community would be to describe the history of several twentieth century sciences. I present instead a cross-section of this history in the form of case studies that indicate the appeal his approach had for particular communities. This account, nevertheless, takes into consideration the subtleties, prejudices, and contradictions inherent in intellectual negotiations. For example, when attempting to integrate Vinogradskii's perspective and methods, even the most exuberant of his supporters were constrained by the demands of their own research agenda, institutional priorities, and personal tastes.

The history of Vinogradskii's influence on science is divided into two parts by the Russian revolution. He did not develop what might be called a "school." His few formal students did become powerful scientists in their own right, yet they established no intellectual lineage in Vinogradskii's name. The group he influenced is best characterized as a "following"—scientists who consistently monitored his publications or admired him as a scientific personality. Some of his followers were primarily influenced by his research in the 1880s–1910s on iron bacteria or sulphur bacteria. Had he remained in retirement, his place in the history of science would have been limited to his development of the elective culture method and his discovery of autotrophism. These dramatic inventions inspired a generation of microbiologists to study microbial nutrition, taxonomy, and physiology. These scientists—from many countries—produced an expansive literature on specific types of microbes (including, but not limited to nitrogen, sulphur, and iron bacteria) and on the general phenomenon of microbial physiology during the twentieth century.

Vinogradskii influenced science much most significantly with the work he conducted during his Brie-Comte-Robert period, particularly through his criticisms of soil microbiology in the 1920s–1940s, the spread of his direct ecological method, and the publication of his 1949 compendium. His ecological methods of the 1920s–1940s attracted a new audience. This audience included many of his earlier supporters who (for some after suffering Vinogradskii's acerbic criticisms) reacted positively to his

[3] For a discussion of the "factors in the emergence of twentieth century biology," see Garland Allen, *Life Science in the Twentieth Century* (Cambridge: Cambridge University Press, 1981), xv–xix. In the 1880s–1920s, life scientists—microbiologists included—were struggling for the same recognition society was lavishing on the physicists and chemists and they sought to meet the ideals of the new kind of knowledge in science. In part, they modeled their approaches on the tenets of these so-called hard sciences. Vinogradskii's early work represents one example of this trend. In the 1920s, however, a new movement arose in biology; life scientists began to reject simple reductionism for integrative approaches.

novel ecological perspective. New adherents in diverse fields drew on Vinogradskii's powerful authority as a "founding father" of their science to promote either their own ideas or new scientific disciplines.[4]

A number of leading figures later known for their application of ecological thinking to a variety of questions appropriated Vinogradskii's scientific approach. Here I will discuss four settings where his methods were incorporated to great effect—at the Agricultural Experiment Stations at Rutgers University, New Jersey and at Rothamsted, England; the Delft School of Microbiology in Amsterdam; and the Department of Agricultural Microbiology, Leningrad. In each case, Vinogradskii's methods for investigating the soil ecologically found an audience among researchers who assimilated his methods into their investigations according to the demands of their own experimental programs.

The "Cycle of Life" at the Rutgers Agricultural Experiment Station

On the banks of the "Old Raritan" at Rutgers University in New Jersey, three generations of agricultural microbiologists drew upon Vinogradskii's work in their studies of soil microbes. At the turn of the century, Jacob Lipman established at Rutgers a program in agricultural microbiology that attracted students from around the world. Among their number was Selman Waksman, the soil scientist who later won the Nobel Prize for his discovery of the antibiotic streptomycin. Under Lipman's direction, Waksman conducted experiments based on Vinogradskii's classical nitrification papers. Waksman himself trained numerous students in agricultural microbiology, including Rene Dubos who developed the first "antibiotic" from soil microbes. Each of these scientists was entranced by Vinogradskii's vision of the living soil and the cycle of life and encouraged by his practical methods to pursue that vision in the laboratory and field.[5] (Waksman considered his debt to Vinogradskii so profound that he wrote a biography of him.)

---

[4] Jan Sapp discussed the issues related to the construction of "new lives" for scientists who become recognized authority figures. See Jan Sapp, "The Nine Lives of Gregor Mendel," *Experimental Inquiries*, ed. H. E. Le Grand, (Dordrecht; Boston: Kluwer Academic Publishers, 1990), 137–166. Vinogradskii found a receptive audience for his ideas across a broad international community. He corresponded with scientists from locations as far away as Shanghai, Brazil, Moscow, and California. See also Krementsov, *Stalinist Science*.

[5] Waksman characterized Lipman's place in the history of soil science as "always . . . connected with the study of the cycle of nitrogen in nature." Lipman had begun his scientific activities at a time when "the role of [microorganisms] in the cycle of nitrogen in nature" was attracting attention. Lipman represented, Waksman thought, the school of thought of Vinogradskii and Beijerinck in America. Selman A. Waksman, "Jacob G. Lipman as an Investigator: A Chapter in the History of Soil Microbiology," in *Soil Science,* Vol. 40, June–December 1935, 11–23, esp. 11.

Jacob Lipman (1874–1939)—a Russian émigré—entered Rutgers University in 1894 and soon devoted himself to agricultural studies.[6] He could hardly had avoided it. He had been admitted to Rutgers through the support of H. L. Sabsovich—an agricultural chemist from Odessa working at the Colorado Agriculturalist Experiment Station.[7] The director of the station, Edward B. Voorhees, took a personal interest in Lipman and encouraged him to study soil chemistry, plant nutrition and "especially the role of microbes in soil processes and plant growth."[8] Lipman pursued these subjects at Cornell, where he earned an MA (1901) and Ph.D. (1903). He returned to Rutgers in 1901 as instructor in agricultural chemistry to establish a department of soil chemistry and bacteriology.[9]

By this time, Lipman was well versed in Vinogradskii's contributions to soil microbiology. In particular, Vinogradskii's nitrification investigations of 1893–1894 informed Lipman's approach to investigating the interrelationship between microbial species in nature.[10] In Lipman's doctoral thesis on *Nitrogen Fixing Bacteria*, for example, he interpreted Vinogradskii's discovery of two nitrogen-fixing microorganisms working in tandem as a symbiotic relationship.[11] He applied that concept not only in his graduate research, but also to fulfill the project Voorhees assigned him in 1901. The German agronomist Paul Wagner's claim that bacteria might cause nitrogen loss in the soil had attracted the worried attention of Voorhees, who charged Lipman to investigate the role of bacteria in the formation, transformation, and destruction of fertilizers.[12]

In their extensive survey of the literature on soil bacteriology (1906), Voorhees and Lipman stressed the central role bacteria played in soil fertility. The struggle for existence, they wrote, had produced "innumerable bacterial species" among which

---

[6] On Lipman see Selman A. Waksman, *Jacob G. Lipman: Agricultural Scientist and Humanitarian* (New Brunswick: Rutgers University Press, 1966) and several articles by Waksman, E. J. Russell, A. W. Blair, and Robert Allison dedicated to Lipman in *Soil Science,* vol. 40, June–December 1935. Interestingly, Vinogradskii, most liekly having been solicited to do so, contributed an article to this volume, wherein he discussed his direct method for the first time in English; Serge Winogradsky, "The Method in Soil Microbiology as Illustrated by Studies on *Azotobacter* and the Nitrifying Organisms," Idem., 59–76.

[7] Sabsovich who directed a program affiliated with the Woodbine colony in Cape May County, New Jersey, noticed Lipman and helped him gain admittance to Rutgers University in 1894. Lipman had emigrated from Russia in 1888, to escape the persecution Jews were experiencing under Tsar Alexander III's regime. The Lipman family found refuge in the Woodbine community, which had been formed with the support of American philanthropists concerned with assisting Russian and Romanian Jews. One of their principle projects was to form agricultural settlements in the United States, and Argentina; see Selman A. Waksman, *Jacob G. Lipman: Agricultural Scientist and Humanitarian* (New Brunswick: Rutgers University Press, 1966), 8–15.

[8] Ibid., 20.

[9] Ibid., 27.

[10] J. G. Lipman, "The Fixation of Atmospheric Nitrogen by Bacteria," *Proceeding of the Twentieth Annual Convention of the Association of Official Agricultural Chemists* (Washington: Government Printing Office, 1904), 146–162.

[11] J. G. Lipman, *Nitrogen Fixing Bacteria* (Cornell University, Doctoral Thesis, 1903), 149–151, 156.

[12] Waksman, *Jacob G. Lipman: Agricultural Scientist and Humanitarian*, 31.

"we find some of our staunchest allies in crop production."[13] Some of the most significant of these organisms were Vinogradskii's sulphur, iron, and nitrogen bacteria, which "play an essential part in the cycle of transformation to which nitrogen, carbon, hydrogen, and sulphur are subject."[14] Vinogradskii's sulphur bacteria investigations had shown them that sulphur bacteria—by virtue of their physiological ability to change hydrogen sulfide into sulfates—played a significant role in "the great cycle of transformation of matter in nature."[15] In 1911, Lipman succeeded Voorhees as director of the experiment station; and by 1915, he dominated all aspects of agricultural studies at Rutgers.[16]

No one celebrated Vinogradskii's role in the development of soil microbiology in the West more than did Lipman's student Selman Waksman. Soon after arriving in the United States from his home in Priluka near Kiev, Ukraine, Waksman visited Rutgers College and met Lipman. Lipman recommended that Waksman pursue his interest in studying the "chemical processes of living systems" not by attending medical school, but rather by following an agricultural curriculum.[17]

Waksman's path to Vinogradskii's ideas was plowed by his farming experiences and Lipman's teaching. Waksman's cousin Mendel Kornblatt—on whose farm he had lived and worked—initiated him into "the mysteries of plant and animal life." Kronblatt gave Waksman his first systematic training in agriculture, instilling in him a "desire to learn the fundamental principles, and the chemical and biological mechanisms that make agriculture possible."[18] According to Waksman's autobiography, this connection between practical agriculture and microbiology led him to soil microbiology. To understand how nature worked, he thought, he needed to learn the numerous chemical and microbiological processes that occur in the soil and "result in the liberation of a continuous stream of nutrients which make possible the continuity of life and which serve to complete the chain of living reactions in nature."[19] He turned to the soil, then, to investigate "the cycle of life in nature."[20]

Waksman first developed an interest in Vinogradskii's soil microbiology during his apprenticeship under Lipman. Beginning in 1913, as part of Lipman's program of lectures and seminars, Waksman attended his first bacteriology course. He finally had found an advisor who could help him understand Kornblatt's practical agricultural lessons and the cycle of life concept on a scientific basis. Waksman adopted Vinogradskii and Lipman's view that soil microorganisms "play an important role

---

[13] Edward B. Voorhees and Jacob. G. Lipman, *A Review of Investigations in Soil Bacteriology*, a U.S. Department of Agriculture, Office of Experiment Stations—Bulletin 194 (Washington: Government Printing Office, 1907), 6–7.

[14] Ibid. 7.

[15] Ibid., 33

[16] Waksman, *Jacob G. Lipman: Agricultural Scientist and Humanitarian*, 31.

[17] Selman A. Waksman, *My Life with the Microbes* (New York: Simon and Schuster, Inc, 1954), 68.

[18] Ibid., 74

[19] Ibid.

[20] Ibid.

in the cycle of life in nature" and studied how to apply that vision to soil science.[21] Waksman found Lipman's devotion to Vinogradskii's work quite natural. Lipman, an "enthusiastic teacher . . . who devoted most of his own life to the study of the function of bacteria in the transformation of nitrogen in the soil, came early under the influence of Winogradsky's work on nitrification and nitrogen fixation."[22] From 1914 to 1915, while still an undergraduate, Waksman joined a small group of graduate students for Lipman's weekly seminar on soil bacteriology—a field they considered Vinogradskii to have "made famous."[23] Waksman recalled that their seminar discussions had the side effect of making them curious about Vinogradskii's fate—he had quietly "disappeared from the scientific horizon" nearly a decade previously. Waksman would solve the mystery in 5 years.

Waksman explored this mystery through his European contacts. In the fall of 1922—now a scientist in his own right—he received an offprint of the first article Vinogradskii published after his emigration from Russia.[24] The arrival of this article from Belgrade was most likely not as surprising as Waksman claimed in his autobiography—for several years he had been corresponding with Omelianskii and other Russian microbiologists, with whom he discussed Vinogradskii. In December 1922, encouraged by Omelianskii, Waksman contacted Vinogradskii at Brie-Compte-Robert.[25] Waksman's letter arrived at what was probably a sensitive time for Vinogradskii—his decision to not emigrate in 1891 had turned out to be only a postponement, and he was struggling to re-launch his career. He no doubt welcomed Waksman's praise and attention, and especially the generous offer to send American reprints and the journal *Science*.[26]

---

[21] Selman A. Waksman, *Sergei N. Winogradsky: His Life and Work, The Story of a Great Bacteriologist* (New Brunswick: Rutgers University Press, 1953), v. Waksman's interest in Vinogradskii had several dimensions: he admired him as a basic scientist, as someone who shared his goal of setting soil microbiology on a firm methodological foundation, and—like Lipman—as a fellow Russian Ukrainian.

[22] Ibid.

[23] Ibid.

[24] The publication was Serge Winogradsky, "Eisenbakterien als Anorgoxydanten," *Zentralblatt für Bakteriologie*, Vol. 57, 1922. The proximate times of publication and Waksman's receipt makes one question Waksman's timeline. His correspondence reveals a slightly different chain of events leading to his eventual contact with Vinogradskii. Waksman found this article interesting enough to present in place of his already prepared seminar paper. He recalled that in it "Winogradsky refuted all the accumulated attacks upon his work, some of which were due to misunderstanding and confusion, that had been made during the nearly two decades of his absence from the scientific field." Waksman, *Sergei N. Winogradsky: His Life and Work, The Story of a Great Bacteriologist,* vi.

[25] Letter from Selman Waksman to Vinogradskii, December 12, 1922, Archiv Rossiskoi Akademii Nauk,, Moskovskii filial, fond 1601, opis' 1, delo 127, list 1. Waksman had learned from a Professor Barthel that Vinogradskii was in Belgrade and then from Omelianskii that he had moved to the Pasteur Institute. Ibid. Waksman also discusses this episode in his, *My Life with the Microbes* (New York: Simon and Schuster, Inc, 1954), 113–114.

[26] Ibid. The dialectic of influence that colors this first letter would be a constant throughout the relationship between Waksman and Vinogradskii. Reading their correspondence and investigative pathways reveals a generous reciprocity of ideas, material goods, and authority. Both benefited from the relationship.

Starting in 1918, Waksman initiated a reform of soil bacteriology at Rutgers. Appointed lecturer in soil microbiology—after studying for 2 years at the University of California, Berkeley—he reorganized the Department of Soil Bacteriology to reflect his own research interests.[27] "Soil Bacteriology" had never encompassed his primary subjects of investigation—microscopic soil fungi and actinomycetes—and he renamed it "Soil Microbiology." Behind this simple name-change lay his growing interest in developing a synthetic approach to studying the soil's microflora.

At a crucial time in this reform process, Waksman visited Vinogradskii at Brie-Comte-Robert. During his 1923–1924 international tour of 50 microbiological laboratories, Waksman observed first hand Vinogradskii's approach to soil microbiology.[28] Its originality and comprehensiveness inspired him to enlist in the old master's effort to create a new discipline of soil microbiology.[29] Waksman attended the May conference in Rome, where Vinogradskii outlined his vision to the International Congress of Soil Science. It must have been quite invigorating for Waksman to hear Vinogradskii deliver such a strong and clear message that coincided with his own reform agenda. Waksman's awareness that he shared the same roots not only with Lipman—who also attended the congress and was elected president of the society—but also with Vinogradskii must have increased the significance and impact of this event on him.

It is striking that Vinogradskii, Waksman, and Lipman were from the same part of the world—Podolia, Ukraine. It is all the more remarkable that many of the prominent Russian and Soviet soil scientists also had roots in this region in southern Russia. Was there something about Podolia that encouraged an interest in the scientific study of the soil? One clue comes from Vasilii Omelianskii, Vinogradskii's student and one of his closest Russian colleagues for over 30 years, and, also from southern Russia. In a 1927 rough draft of his "New Paths in the Development of Soil Microbiology," he described "the soil, from which everything comes, and into which everything inescapably returns" as having "something about it that attracts all of us, whether the poet or the state."[30] For Russians, "with their elemental (*stikhiinoi*) love for the earth-benefactress (*zemlia-kormiletsa*), the question of the earth presents a completely special, exceptional interest."[31] For Omelianskii, Russians could not escape the legacy of the soil: "The maintenance of the soil—though in most cases, it was, alas, done entirely irrationally—consumed the entire working life of the land

[27] Upon return from his California trip, Waksman found that his assistants had joined the Army, his laboratory in ruin, and all his laboratory cultures long dead. Ibid., 102.

[28] Letter from Waksman to Vinogradskii, September 19, 1924, fond 1601, opis 1, delo 127, list 2. He sent the letter from Paris, which means that he most likely had accompanied Lipman to the International Conference of Soil Science where Vinogradskii presented his direct method.

[29] Letter from Waksman to Vinogradskii, November 29, 1924, fond 1601, opis 1, delo 127, list 3. Waksman published his first statement along these lines in a review of his travel experiences and on the currently poor state of soil microbiology.

[30] V. L. Omelianskii, a draft of his *Novye puti v razvitii pochvennoi mikrobiologii*, Archiv Rossiiskoi Akademii Nauk, Peterburgskii Filial, fond 892, opis' 1, delo 8, listy 1–2.

[31] Ibid.

workers (*zemlerod*). A good harvest determined the prosperity of the peasants, and the prosperity of the peasants dictated the ability to feed the entire country. The soil held more significance for those, like Omelianskii, who were raised in the "bread-basket" of Russia, than for those born in urban settings.[32]

After his European trip in 1934, Waksman wrote a warm letter to Vinogradskii announcing him his "actual teacher."[33] From the very beginning of Waksman's 10 years investigating soil microorganisms, he wrote, "he had repeatedly returned to Vinogradskii's classic work in their science."[34] This virtual apprenticeship with Vinogradskii and the force of his personality and scientific authority, led Waksman to adopt Vinogradskii's vision for soil microbiology. At the time of their meeting, Vinogradskii was proselytizing for a new discipline of soil microbiology—an endeavor Waksman readily supported. The old master and the fresh adept, they were both convinced of the need "to place their science in its needed place—on a foundation of pure science."[35]

Waksman extended Vinogradskii's influence in many ways. He applied Vinogradskii's methods in his own research, participated in Vinogradskii's expanding scientific network, assisted in publishing Vinogradskii's 1949 collected works, and wrote a laudatory biography of him. His most enduring contribution to Vinogradskii's legacy, however, was through his many students.[36] Following Lipman's example, Waksman stimulated his students by placing them in a venerated tradition in the history of microbiology—they were the descendents of the Pasteur-Vinogradskii lineage. In the mid 1940s, for example, Waksman's student and self-proclaimed disciple, Jackson Foster, found his University of Texas students "starving for [a]

---

[32] It is significant that the soil scientists at Rutgers, and Omelianskii and Vinogradskii were all southern Russians. Perhaps it has to do with the network of Russian émigrés in the United States, who shared familial contacts and, in this case, an interest in science and agriculture. Originating in the same area as Vinogradskii, they had imbibed a worldview amenable with the cycle of life concept. As they built their institutions, research programs, and wrote histories of their disciplines, this concept provided one unifying principle. Southern Russia was not the epicenter of the cycle of life concept—it has a long history in ancient mythology, folklore, religion, and science. These southern Russians, however, were some of the first to develop the cycle of life concept into a scientific agriculture movement.

[33] Letter from Waksman to Vinogradskii, November 29, 1924, Archiv Rossiiskoi Akademii Nauk, Peterburgskii Filial, fond 1601, opis 1, delo 127, list 3.

[34] Ibid.

[35] Their mutual interest in this project did not prevent them from vigorously debating the merit of various laboratory methods over the next three decades. Ibid.

[36] Waksman spent nearly his entire scientific career at Rutgers, where he fostered new generations of soil microbiologists. This successful career culminated in the founding of his internationally renowned Institute of Microbiology in the 1950s. Even at the dedication of this institute Waksman recalled his debt to the cycle of life concept when he asked: "How did this all come about? How did the humble searcher for microbes in the earth under our feet and in the seas around us, one whose primary concern was the role of infinitesimal living things in the complex cycle of life on this planet, succeed in bringing this about?" Selman A. Waksman, *My Life with the Microbes*, ix.

microbiology" based on the "essential fundamentals, e.g. . . . autotrophism, Winogradsky, Waksman, [and] applied microbiology."[37] Foster credited Waksman with being first to make "it possible for Winogradsky's peers and successors to really appreciate him."[38]

Waksman also spread Vinogradskii's message to a large American and international audience.[39] Writing the first comprehensive textbook and general guide to soil microbiology, Waksman made clear his commitment to Vinogradskii's approach and cycle of life perspective. Waksman organized his book by Vinogradskii's fundamental principle that soil microbiology must be based, not on the study of isolated microorganisms, but rather on their activities and role in their natural environment. Guided by Vinogradskii's ecological vision, Waksman synthesized in his nearly 900-page tome 2,500 Russian, German, French, and English publications on the soil sciences. The clear, central message was Vinogradskii's: soil microbiologists, first and foremost, needed to investigate microbial activity ecologically; that is, as a biological process influenced by the soil's (or water's) physical and chemical properties.

A "Holistic Habit of Mind"[40]

One of Waksman's first students, René Dubos, arrived at Rutgers already familiar with Vinogradskii's ecological views. Dubos, who earned his Ph.D. in soil microbiology under Waksman, recalled that he conducted his doctoral work in the 'spirit of Vinogradskii.'[41] Like Vinogradskii and Waksman, Dubos was raised in an agricultural region. After considerable searching, Dubos began a career in scientific agriculture

[37] A letter dated October 19 (unspecified year) from Jackson W. Foster—Associate Professor of Bacteriology, in the Department of Botany and Bacteriology of the University of Texas, Austin—to Selman Waksman, 4. I place the letter close to 1947. From the Selman Waksman Papers, Collections of the Manuscript Division, Library of Congress. Foster was also greatly impressed by Waksman's historical sensibilities and especially by Waksman's recently published biographical article on Vinogradskii in *Soil Science*.

[38] Ibid., 4

[39] Selman A. Waksman, Principles of Soil Microbiology (Baltimore: Williams and Wilkins Company, 1927, 1932). Waksman dedicated the second edition to "the investigators who have thrown the first light upon some of the most important soil processes and whose contributions can well be considered first and foremost in the science of Soil Microbiology"—that is, Beijerinck and Vinogradskii. Ibid.

[40] Gerard Piel used this phrase to describe Dubos' approach to science. Dubos always chose "his experimental target in the context of his big question." Gerard Piel and Osborn Segerberg, Jr, eds., *The World of René Dubos: A Collection o f His Writings* (New York: Henry Holt and Company, 1990), xx.

[41] Carol L. Moberg and Zanvil A. Cohn, "René J. Dubos," *Scientific American* (May 1991), 66–72. René Dubos, "Each a Part of the Whole," Gerard Piel and Osborn Segerberg, Jr., eds., *The World of René Dubos: A Collection of His Writings* (New York: Henry Holt and Company, 1990), 6.

at the Institut National Agronomique in Paris in 1919.[42] His education at "L'Agro" prepared him to think in symbiotic or ecological terms about not only agriculture, but also about human physiology, zoology, and geology. In the chemistry lectures there Dubos first encountered Vinogradskii's work. Dubos' instructor, Gustav Andre, based his course on agricultural chemistry on the works of Boussingault, Schloesing and Müntz, Pasteur and—when discussing soil microorganisms—on Vinogradskii's nitrogen-fixing research.

A series of events led Dubos to Rome for the 1924 International Congress of Soil Science. After graduating in 1921, he continued his studies at the Institute National Agronomique Coloniale. Training there as an *ingéniur agronome* reinforced the holistic perspective Dubos had acquired during his agricultural education. Health problems thwarted his hopes of going to Indochina (a French colony), so instead he found a good job at the International Institute of Agriculture in Rome. While working as associate editor for his institute, he came across an article that changed his life direction yet again.

In February 1924, Vinogradskii published a general description of his direct method in *Chemie et Industrie*, one of the technical journals Dubos abstracted for his Institute's organ.[43] This short article made a powerful impression on Dubos, becoming an integral part of his mature scientific vision and self-image.[44] Dubos later recalled that his reading of Vinogradskii's idea "that microorganisms should be studied not in artificial laboratory cultures but in their natural environments in competition with other bacteria" inspired him to launch a career in microbiology and to embrace an ecological approach to it.[45] It is likely that later that year, Dubos heard Vinogradskii's presentation about his direct method at the International Congress in Rome. Here too, Dubos made contact with Lipman and Waksman. Lipman invited Dubos to come to Rutgers to continue his studies. Waksman reiterated that offer when the two met by chance on the steamer to America.[46]

Dubos applied Vinogradskii's perception that "countless microbes perform limited, well-defined tasks to recycle organic matter so that it does not accumulate in nature" to his investigation at Rutgers and it became the core of his enduring scientific vision.[47] In Waksman's lab, Dubos investigated actinomycetes in the competitive environments of their natural soil habitats. Here he applied an ecological approach to cellulose decomposition in the soil (a subject to which Vinogradskii had devoted

[42] Jill Elaine Cooper, *Of Microbes and Men: A Scientific Biography of René Jules Dubos*, Ph.D. Dissertation (Rutgers, The State University of New Jersey, 1998), 28–31.

[43] Serge Winogradsky, Sur la méthode directe dans l'étude microscopique du sol, *Chemie et Industrie*, Vol. 11, No. 2, February 1924.

[44] He recounted the influence Vinogradskii had on him during his oral interview with Saul Benison in 1955–56. See Jill Elaine Cooper, *Of Microbes and Men: A Scientific Biography of René Jules Dubos*, 13, (footnote 4).

[45] Carol L. Moberg and Zanvil A. Cohn, "René J. Dubos," *Scientific American* (May 1991), 66.

[46] Gerard Piel and Osborn Segerberg, Jr, eds., *The World of René Dubos: A Collection o f His Writings* (New York: Henry Holt and Company, 1990), 6.

[47] Moberg, "Rene J. Dubos," 66–67.

considerable time). When he presented the results of this laboratory research at the 1927 International Congress of Soil Science in Washington, Dubos felt that his underlying ecological message had resounded with his audience. Had it been obscured by the details of his report, his message was amplified by the chair of the session, English agricultural bacteriologist Sir John Russell, who "brought out in a clear manner the ecological significance of [Dubos'] findings."[48] Expecting to publish his report in the *Journal of Bacteriology*, Dubos instead accepted the surprise request to publish it in *Ecology*.[49]

In 1927, when Dubos shifted his interests from soil microbiology to medical bacteriology, he brought Vinogradskii's ecological perspective with him. At that time, even the driven reformer Waksman had to admit that the discipline of soil microbiology offered few career opportunities to recent graduates like Dubos.[50] After completing his dissertation, Dubos eventually found a new position in Oswald Avery's laboratory at the Rockefeller Institute for Medical Research. Dubos approached the central question of Avery's investigation of pneumococcal pneumonia—how to decompose its protective polysaccharide envelope—on Vinogradskii's terms. Just as Vinogradskii (following Jean-Baptiste Dumas, Pasteur, Famintsyn, and Cohn) had espoused in his speech on the cycle of life in 1897, Dubos explained to Avery that "if there were no enzyme that could decompose that capsular polysaccharide, it would accumulate in nature, there would be mountains of it now. So there must be, somewhere in nature, some microbe that would decompose it."[51] Within 2 years, using a version of the elective culture method, he had found that microbe. His discovery of the enzyme this organism produced (he named it ribonuclease) provided one of the foundations for developing antibiotics.[52]

His commitment to Vinogradskii's approach grew even stronger over subsequent years. By 1954, he summed up his approach with the axiom "that any metabolic analysis of the infectious process must be placed on an ecological basis."[53] Medical microbiology would benefit, he recommended, from the approach Vinogradskii had

---

[48] Saul Benison, "René Dubos and the Capsular Polysaccharide of Pneumococcus: An Oral History Memoir" *Bulletin of the History of Medicine*, Vol. 50, No. 4, Winter 1976, 459–477; see Dubos' comments on 462.

[49] Dubos recalled that he did not even know of the journal's existence when the editor approached him. Ibid., 462. Although, any of the 16 members of the editorial board might have contacted Dubos, two likely suspects are its editor-in-chief Barrington Moore whose interest in soil science is clear in his opening comments in the first issue of *Ecology*; and Lipman's brother, Charles B. Lipman, who participated in Vinogradskii's network.

[50] For a discussion of this topic, see Jill Elaine Cooper, *Of Microbes and Men: A Scientific Biography of René Jules Dubos*, 83–86.

[51] Saul Benison, "René Dubos and the Capsular Polysaccharide of Pneumococcus: An Oral History Memoir," 464.

[52] Moberg, "René J. Dubos," 70. It is ironic that the scientist to make some of the most major contributions to this new science was Dubos' previous mentor, Selman Waksman, who won a Nobel Prize for his research.

[53] Dubos made these points discussing "The fate of Microorganisms in Vivo," in René J. Dubos, *The Biochemical Determinants of Microbial Disease* (Harvard University Press, 1954), 22.

formulated for soil microbiology in his early Brie-Compte years. Dubos reminded medical bacteriologists of Vinogradskii's warning that artificial culture media could offer no conclusive evidence about microorganisms in soil processes. Like Vinogradskii, they should develop techniques for studying "microbial activities in an environment as similar as possible to that found in the soil."[54] For Dubos, Vinogradskii's ecological conceptualization of the soil applied also to the body and infectious diseases. He translated Vinogradskii's idea that "each fragment of the soil constitutes a biosphere with its own chemical and biological peculiarities" into a vision for studying disease in vivo, wherein that soil is replaced by the body.[55]

Dubos's influence extended increasingly beyond the realm of medical bacteriology and soil science. Beginning in the 1950s, he shifted his interests from medical microbiology to human ecology. By the 1970s, he had became "an elder statesman of the environmental movement" translating his complex ideas into popular slogans such as "Think globally, act locally."[56] Through his extensive lecture tours and his numerous books and articles for the educated layperson, Dubos transformed Vinogradskii's vision of the cycle of life into a broad environmentalist and ecological perspective in the mid-late twentieth century.[57]

Statistical Soil Science: Rothamsted Agricultural Experiment Station

While Vinogradskii was "recapturing" his investigations in France in the 1920s, Sir John Russell was fighting to rejuvenate the Rothamsted Agricultural Station in England. Formed in 1848 in Harpenden, the Rothamsted Station was one of the world's oldest scientific agricultural institutions, having served as a place for botanists, bacteriologists, and chemists to investigate fertilizer usage, crop production, and animal nutrition.[58] Serving as the station's new director since 1912, Sir Russell set a new research direction focused on soil chemistry and physics,

---

[54] Ibid., 22–23.

[55] Ibid., 23.

[56] Moberg, "René J. Dubos," 74.

[57] Dubos' mature ecological perspective, expressed in his many slogans, such as "Think Globally, act Locally," continued to reflect the sentiments of Vinogradskii's cycle of life concept.

[58] John Bennet Lawes founded the Lawes Agricultural Trust at Rothamsted Agricultural Experiment Station in 1843, making it the oldest institution of its kind (at least in England). The station expanded continuously until finally the British Government purchased it in 1934. The botanist, bacteriologists, and chemists housed there investigated the use of fertilizers crop production, and animal nutrition. See Sir E. John Russell, *A History of Agricultural Science in Great Briton, 1620–1954* (London: George Allen & Unwin, Ltd., 1966), 289–332; Edward J. Russell, *Soil Conditions and Plant Growth* in the series *The Rothamsted Monographs on Agricultural Science* (London: Longmans, Green and Co., 1921); Idem., *The Microorganisms of the Soil* (Longmans, Green, and Co., 1923); and Idem., *The Land Called Me; An Autobiography* (George Allen & Unwin, Ltd., 1956);

microbiology, and the statistical analysis of data. In this newly reorganized institution, Sir Russell, his colleague H. G. Thornton, and their students (especially Ward Cutler) investigated the relationship between soil conditions and plant growth. During this reformative period at Rothamsted, the Russell group engaged Vinogradskii at international congresses and in correspondence. In the 1920s–1940s, they assimilated his ecological perspective and direct methods in their ongoing investigations of soil microorganisms.

They came to know each other through a correspondence that began as an exchange of compliments and soon turned to an exchange of techniques, soils, and ideas.[59] After the Rome International Conference of Soil Science in 1924, Russell sent a French translation of his *Soil Conditions and Plant Growth* to Vinogradskii, who had cited it in his recent nitrification articles. Russell, who was very familiar with Vinogradskii's earlier work, addressed Vinogradskii as "one of the founders of [soil microbiology]."[60] After the Rome conference, moreover, Russell wanted to integrate the direct method into his laboratory's approach and he sought out Vinogradskii for suggestions to improve their research.[61] During this period, Russell and Vinogradskii shared an interest in researching how to use microbiology to improve arable lands.[62] Through this dialogue, Russell and his students kept in close touch with developments in Vinogradskii's direct microscopic examinations of the soil.

Vinogradskii readily suggested improvements to the Russell group, who listened to his advice attentively yet hesitantly. Their reluctance to adopt Vinogradskii's complete direct method reflected the difference in scope inherent in the respective programs of study: Vinogradskii investigated the changing relationships among microorganisms in their natural habitats and maintained a global perspective rooted in the cycle of life. The Russell group, however, focused its work more narrowly and pragmatically—investigating how microbes influenced plant (and crop) growth in agricultural plots. This distinction led to disagreements over the proper methods

[59] Vinogradskii first identified his research as ecological in a letter to Russell in 1927. Here he responded to Russell's impression of Vinogradskii's review of Waksman's *Principles of Soil Microbiology*. In the review, Vinogradskii stresses that "an ecological spirit is necessary in [soil microbiology] in the same way as in botanical or zoological studies, which leads to disengaging (*a dégager*), as much as possible, the play of natural forces from the effects of human interference." Serge Winogradsky, "Revue Critique: Principes de Microbiologie du Sol," *Annales de l'Institut Pasteur*, Vol. 41, No. 10, Octobre 1927, 1126–1138, see 1128–1129 for quote. For the letter see the Winogradsky Papers, Service des Archives, Institut Pasteur, Box, Win 2, Correspondance International, Angleterre, Russell folder.

[60] Russell to Vinogradskii, 21 October 1924. Russell also requested Vinogradskii's portrait, which would "take a high place in [his] collection." Ibid.

[61] Ibid.

[62] In 1924, Vinogradskii published two reports on the microbiology of arable soils: Serge Winogradsky, "Sur la microflore autochtone de la terre arable," *Comptes Rendus hebdomadaires des Sciences de l'Académie des Sciences*, 1924, Vol. 178, 1236–1239; and Idem., "Sur l'étude de l'anaérobiose dans la terre arable," *Comptes Rendus hebdomadaires des Sciences de l'Académie des Sciences*, Vol. 179, 861–863.

for studying microbial life and over the most essential measurement criteria. The ultimate goal in the Russell group's investigations, for example, was the determination of fluctuations in microbial populations, which they assessed using plating methods. The plating method was in essence a pure culture method and it drew harsh criticism from the Russell group's newest advisor in Brie-Compte-Robert.

Even though Vinogradskii himself included pure cultures as part of his direct method, he felt justified in criticizing the Russell group's approach. The Russell group, he thought, relied too heavily on plating methods for making their conclusions about the role of microbes in the soil.[63] On the one hand, Vinogradskii allowed that pure cultures were "the best method of studying *all* physiological *possibilities* of the [micro]organisms and their reactions to external conditions."[64] These methods, however, failed to satisfy Vinogradskii's own goal: to study the microbes' "*actualities*" in a natural medium—the real aim of soil microbiology—the artificial method was unreliable.[65] Soil microbiologists should instead use a method that uncovered the free play of microbial activities under conditions as close as possible to the natural ones, that is, to the natural state of the soil.[66] Vinogradskii argued that the Rothamsted group, by basing its approach on pure cultures, was incorrectly assuming that soil microorganisms produced the same biochemical processes in laboratory media as they did in the soil.[67]

As much as he respected Vinogradskii's great authority, Russell defended his group's methods. He contended that they had received good estimates of fluctuations in bacterial number and that those numbers expressed something important about soil conditions such as its fertility. Reacting strongly to Russell's contention that "a separate study of soil microbiology is a mere matter of convenience," Vinogradskii insisted that "a fundamental distinction [existed] between general microbiology—the study of microscopic morphology and physiology—and soil microbiology, which is properly speaking microbial ecology."[68]

If Vinogradskii failed to convince Russell to rely solely on the direct method, he fared better with Thornton, head of Rothamsted's bacteriological department. Thorton attempted to synthesize the Rothamsted lab's methods of bacterial statistics with Vinogradskii's direct method. Russell shared his letters with his colleagues,

---

[63] The Russell group relied heavily on pure cultures to investigate microbial physiology, tracking changes in these cultures with bacterial counts and statistics. During this period, R. A. Fisher assisted the Russell group with their statistical modeling. Vinogradskii assessed their plate methods in his review of Waksman's *Principles of Soil Microbiology*, a copy of which he sent Russell in 1927. Serge Winogradsky, "Revue Critique: Principes de Microbiologie du Sol," *Annales de l'Institut Pasteur*, Vol. 41, No. 10, Octobre 1927, 1126–1138, see 1128–1129 for quote. For the letter see the Winogradsky Papers, Service des Archives, Institut Pasteur, Box, Win 2, Correspondance International, Angleterre, Russell folder.

[64] Vinogradskii to Russell, undated response to Russell's letter of November 28, 1927, 1.

[65] Ibid.

[66] Ibid.

[67] Ibid.

[68] Ibid., 2.

leading Thornton to initiate his own correspondence with Vinogradskii. Thornton had followed "with great interest" the correspondence between Russell and Vinogradskii on the methods of investigating soil bacteria.[69] Vinogradskii's work on the direct examination of these organisms interested Thornton "extremely" and he entirely agreed with Vinogradskii's "opinion that the Ecology of bacteria in soil can be studied most directly by such a method."[70] Thornton was troubled, however, by the difficulties he perceived in using a method that allowed one "to distinguish morphological groups" of bacteria and not "the physiological groups" that were "of importance in relating the activity of bacteria to biochemical processes in the soil."[71] He recommended supplementing the observations made with Vinogradskii's direct method "by studying the bacteria they [the Rothamsted group] had found to multiply in the soil in isolation."[72]

Like Russell, Thornton took issue with Vinogradskii's criticism of their statistical measurement of bacterial population fluctuations. Agreeing with Vinogradskii that they needed to compare their results with "total counts by the direct method," he pointed out, however, that their studies had shown that different groups of soil bacteria fluctuated at different rates.[73] Demonstrating his rather naive understanding of the direct method, Thornton suggested supplementing it with experiments using an array of "differential media."[74] Thus supplemented the new approach would alleviate what he considered the inadequacies of both the direct method and his own plate methods, that is, it would be flexible enough to segregate the varied fluctuations of different groups of soil bacteria.[75]

Vinogradskii's reaction to Thornton's suggestion caught him off guard. Thornton thought that incorporating a technique of differential media in the direct method might help in analyzing the composition of the bacterial flora.[76] Vinogradskii responded with a description of his own, sophisticated versions of the differential media technique. He told Thornton that studying soil samples in pure cultures—ostensibly, Thornton's plate method—could only provide an indication of the density of bacterial populations in the soil.[77] "In order to study their activity," Vinogradskii explained,

---

[69] Undated letter from Thornton to Vinogradskii, probably written in late 1927-early 1928, 1. Winogradsky Papers, Service des Archives, Institut Pasteur, Box Win 2, Correspondance International, Angleterre, Thornton Folder.

[70] Ibid.

[71] Ibid.

[72] Ibid.

[73] Ibid.

[74] Ibid. It is unclear why Thornton did not immediately correlate his differential medium with Vinogradskii's spontaneous cultures. One possibility is that he may not have read Vinogradskii's French *Annales* reports carefully enough and, thus, accessed the direct method only through the filter of Russell's interpretation.

[75] Ibid. The inadequacy of the plate method had forced the Russell group to employ "a differential medium upon which the fluctuations . . . do not cancel one another." Ibid.

[76] Ibid.

[77] Draft of Vinogradskii's response to Thornton, around 1927–1928, 2. Winogradsky Papers, Service des Archives, Institut Pasteur, Box Win 2, Correspondence International, Anglettere, Thornton Folder.

Thornton needed to perform two steps of his direct method. Referring Thornton to his recent *Annales* article, Vinogradskii provided instructions for using spontaneous cultures and elective cultures on silica-gel. These Vinogradskii thought, would allow Thornton and his colleagues to analyze nature as he had in Brie-Comte-Robert. By using elective environments, they could not only identify the varied bacterial groups (species) with their differential population fluctuations, but also correlate these fluctuations with the chemical make-up of their environment. Thornton took the advice to heart and began to try the new methods. By 1934, he was sufficiently convinced to plan a trip to Brie-Comte-Robert to learn more.[78]

Examining the Russell group's debate over the value of Vinogradskii's direct method for their own projects provides a better understanding of Vinogradskii's legacy. At the core of Vinogradskii's critique lay a vision of reforming soil science through the application of novel general methods. Through a correspondence that was initiated in the glow of veneration—but quickly turned to practical matters and the sensitive issues of research preferences—Vinogradskii promoted his perspectives and methods to one of the most important international schools of soil science. In addition, as with Dubos, the particular interests and backgrounds of the Russell group led it to use Vinogradskii's methods in a particular way—perhaps not totally satisfactory to Vinogradskii, but nevertheless incorporating into their work his fundamental insights.

In Beijerinck's Backyard: The Delft School of Microbiology

Vinogradskii's ecological perspective also found adherents at the school founded by one of his closest scientific competitors. The Delft school of microbiology, founded in 1895 by Martianus Beijerinck—who had raced Vinogradskii to an understanding of nitrification—produced a student who would contribute greatly to Vinogradskii's influence on ecological microbiology Lars-Gunnar Romell. The director of the institute, Albert J. Kluyver, who succeeded Beijerinck in 1922, trained Romell in what Kluyver called the Beijerinck-Winogradsky approach.[79]

[78] Letter from Thornton to Vinogradskii, 15 September 1934, 2. There is no evidence, however, that Thornton visited Vinogradskii laboratory. Winogradsky Papers, Service des Archives, Institut Pasteur, Box Win 2, Correspondance International, Angleterre, Thornton Folder.

[79] Romell referred to the similarities between the "great founders of soil microbiology" in his article "Winogradsky's Quest for a Method in Soil Microbiology." As different "scientific personalities as they were, Romell noted that they shared a common background in botanical training during the "golden pioneer time of modern botany." That they did not "start as chemists, agronomists, or bacteriologists," he thought, may account for the "wide biological outlook," a preference for "methods based on direct observation," and the technical "ingenuity" that they shared. See Lars-Gunnar Romell, "Winogradsky's Quest for a Method in Soil Microbiology," *Zentralblatt für Bakteriologie, Parasitenkunde und Infektionskrankheiten*, Abt. II, Vol. 93, 1936, 442–448; see 442 for quotes.

Kluyver himself found little opportunity to apply Vinogradskii's direct method in his own research, yet he valued Vinogradskii's contributions over the years.[80] In particular, he respected the ecological methods Vinogradskii developed at Brie-Comte-Robert, in which Kluyver recognized the ideas of his teacher Beijerinck. With Romell acting as intermediary, in 1925 Kluyver invited Vinogradskii to visit the Delft School to speak on his current research.[81] Later, after their relationship had grown closer, Kluyver invited Vinogradskii to participate in the Sixth Botanical Congress in Amsterdam in 1936. Although Vinogradskii was unable to attend, he worked with Romell to write a paper entitled "Winogradsky's Quest for a Method in Soil Microbiology."[82] In a Jubilee Volume of the *Antoine van Leeuwenhoek Journal of Microbiology and Serology*, Vinogradskii's "Principles of Ecological Microbiology" (in French) occupied the honored position of first article.[83] These invitations reflected the mutual respect Kluyver and Vinogradskii held for one another, a respect that they also declared in their scientific publications.

Kluyver paid the highest compliment he could to Vinogradskii by sending one of his students to apprentice at the Brie-Comte-Robert laboratory. Coming under the direct influence of "the old master in his laboratory" in the mid-1920s, Romell embraced Vinogradskii's effort to impart a new ecological direction to microbiology, and he acutely appreciated the role of the direct method in that goal.[84] Prior to his apprenticeship with Vinogradskii, Romell already had adopted an ecological perspective in his scientific research.[85] Through their work together, teacher and student mutually reinforced their ecological convictions: Romell helped Vinogradskii

---

[80] Trained as a chemist, Kluyver focused his scientific energies on studying the biochemistry of microorganismal metabolism. Studying "the chemical activities of microorganisms," he thought, would reveal the advantages of what he called "comparative biochemistry." Albert J. Kluyver, *The Chemical Activities of Micro-Organisms* (London: University of London Press, 1931), 5. On Kluyver see Pieter Smit, "Albert Jan Kluyver," *Dictionary of Scientific Biography*, Vol. 7, 405–407 [Smit lists The Jubilee Volume for Kluyver as a biographical source, which it is not. It is a collection of scientific reports presented in Kluyver's honor.]; C. B. van Niel, "The "Delft School" and the Rise of General Microbiology," *Beijerinck and the Delft School of Microbiology* (Delft: Delft University Press, 1995), xiii–xxvii.

[81] A series of letters between Kluyver and Vinogradskii, January 2, 1925–February 15, 2005. Winogradsky Papers, Service des Archives, Institut Pasteur, Box Win 2, Correspondence International, Pay Bas, Kluyver Folder. Included in this folder is a correspondence concerning Kluyver's assistance in locating Vinogradskii's daughter Katherine Blawdziewicz, who was behind enemy lines in Poland. Kluyver acted as intermediary for the reunited Vinogradskii family. Letters dated October 13, 1939–April 19, 1940. Idem.

[82] See Serge Winogradsky, "Principes de la Microbiologie Oecologique," *Antoine van Leeuwenhoek Journal of Microbiology and Serology*: Jubilee Volume issued in honor of Albert J. Kluyver, Vol. 12, Nos. 1–4, 1947, 1–16; see 16. Romell's article appeared simultaneously in the *Zentralblatt fur Bakteriologie, Parasitenkunde und Infektionskrankheiten*, Abt. II, Vol. 93, 1936, 442–448.

[83] Ibid.

[84] Lars-Gunnar Romell, "Winogradsky's Quest for a Method in Soil Microbiology," 443.

[85] Romell published an article in 1922 on ecological forces at play in humus formation. Lars-Gunnar Romell, "Luftväxlingen i marken som ekologisk faktor," *Meddelanden från Statens Skogsförsöksanstalt* Vol. 19, No. 1, Stockholm.

see the ecological logic of his own research, and Vinogradskii provided Romell a method to study the forest ecologically.[86] Romell had such a positive experience that he became Vinogradskii's fervent disciple, promoting the direct method as a way to transform soil science into an ecological science.

Romell worked actively to promote Vinogradskii's ecological methods and also helped to solidify Vinogradskii's legacy to ecological microbiology. After his apprenticeship with Vinogradskii, he found employment at the New York State College of Agriculture at Cornell University. There he applied the direct method to his research, and by 1935 had integrated Vinogradskii's approach and findings into his investigations of forest soils. On a variety of occasions during the 1930s, Romell attempted to proselytize for the direct method among the uninitiated.

His efforts met with mixed success. In his talk on the "Importance of Microbiological Investigations in the Study of Agricultural Problems" at the Sixth International Botanical Congress convened in Amsterdam in 1935 (and later published later in the *Zentralblatt für Bakteriologie*), Romell summarized the principles underlying Vinogradskii's "quest for a method in soil microbiology."[87] Appearing the same year as Vinogradskii's critical review of methods in soil microbiology in

[86] Vinogradskii influenced most directly the few research assistants he attracted to his Brie-Compte laboratory. Two, Romell and a Polish scientist, Jadwiga Ziemiecka, went on to establish themselves as path breaking researchers in soil microbiology. Ziemiecka worked in Vinogradskii's laboratory during the summers from 1924 to 1927 and published a collaborative work at the end of her apprenticeship in on nitrogen fixation. Vinogradskii and Ziemiecka collaborated to investigate fixation, a soil process the produced a supply of nitrogen in the soil that could be used by plants. Here they challenged the standard method for establishing this feature of the soil. This method was too rudimentary, Vinogradskii thought, to explore subtle questions about soil dynamics; the past methods were insufficient for studying "the activity of a specific population within the soil." In order to study distribution of the species central to fixation in nature and to determine its role there, a method would need to measure the species' density in the environment. Ziemiecka and Vinogradskii applied the direct method to this problem.

In 1927, Ziemiecka left Paris for the United States to serve as delegate of the Polish Ministry of Agriculture delegate to the International Soil Science Congress in Washington. There she reported on her work in Vinogradskii's laboratory and solidified her membership in the international community of soil science. After these trips, she returned to Poznan University in Poland, where she became assistant professor in soil science. Returning to Poland after 1931, Ziemiecka directed the microbiology section at the State Research Institute of Rural Husbandry in Pulavy, which she eventually transformed into one of Poland's most important research centers in soil microbiology. Here she continued the line of investigation she had initiated under Vinogradskii's mentorship on the influence of microorganisms on the soil.

Ziemiecka organized her investigations primarily around Vinogradskii's direct method. She combined the direct method, however, with the slide culture technique of Rossi and Kholodnyi. Vinogradskii expressed his skepticism that their method could say anything about soil dynamics and the microflora ecology. In the face of his criticism, Ziemiecka maintained that the slide culture method was very helpful in studying the influence of the characteristics of the soil and the cultural conditions on the micro-vegetations in the soil. She did not replace any part of the direct method—and only added it to its other techniques.

[87] Lars-Gunnar Romell, "Winogradsky's Quest for a Method in Soil Microbiology," 442–448.

the American journal *Soil Science*, Romell's article helped to introduce Vinogradskii's direct method to a wider audience of soil scientists.[88] In these publications, he stressed the significance of Vinogradskii's direct method for achieving ecological soil microbiology.

In heroic tones, Romell told the story of Vinogradskii's return to soil microbiology in 1922 and of the old master's subsequent critique of current methods.[89] In Romell's opinion, it had not been merely the fact that Vinogradskii had lived to see "leading workers like Beijerinck" obscure the "clear cut results of [Vinogradskii's] classical work on the autotrophic bacteria" that had led him to attack the use of pure cultures.[90] More important than these experiences, Romell thought, was Vinogradskii's "clear recognition of the fact that *soil microbiology is not mere bacterial physiology*, but has to deal with the *ecology of the microbial soil population*."[91] Romell traced the roots of Vinogradskii's ecological approach to his discovery of autotrophic organisms—that is, chemosynthesis—50 years earlier. For Romell, it was this experience that gave Vinogradskii the confidence to adopt an "ecological way of tackling things" in the 1920s and later.[92]

Romell extended Vinogradskii's influence through his forestry research and proselytizing efforts. After working with Vinogradskii in Brie-Comte-Robert, Romell moved to the New York State College of Agriculture in Cornell University where he researched forest soils.[93] At Cornell, Romell "spent some leisure hours . . . outside the main program," as he called it, investigating the organisms responsible for forming nitrates in forest soils.[94] He applied the direct method to identify for the first time a nitrite organism in forest soils.[95] The key to this discovery was Vinogradskii's view that to understand the nature of a microbe, one had to study it in its natural conditions. Romell applied Vinogradskii's methods to study the bacterial life cycle in the forest soil and thus reveal the ongoing biological reactions in this microbial landscape.

---

[88] Ibid. See also Serge Winogradsky, "The Method in Soil Microbiology as Illustrated by Studies on *Azotobacter* and the Nitrifying Organisms," *Soil Science*, Vol. 40, June–December, 1935, 59–76.

[89] Romell, "Winogradsky's Quest for a Method in Soil Microbiology," 443.

[90] Ibid.

[91] Ibid.

[92] Ibid., 443–444. For Romell, the methods Vinogradskii developed in the 1920s made it possible to study natural competition within the soil environment and the reaction of the soil population as a whole. He applied this concept in his own research, in which he investigated the "natural living soil" as a micro-ecological plant community. Letter from Romell to Vinogradskii 20 April 1935, 7. Winogradsky Papers, *Archives de l'Institut Pasteur*, Box WIN 2, L. G. Romell Folder.

[93] Romell was the Charles Lathrop Pack Research Professor in Forest Soils at Cornell from approximately 1929–1935. Then he returned to Sweden where he found a position at the Swedish Forestry Research Institute.

[94] Lars-Gunnar Romell, "A Nitrosocystis from American Forest Soil," *Svensk Botanisk Tidskrift*, Vol. 26, Nos. 1–2, 1932, 303–312; see 306 for quote.

[95] Romell found it easy to identify this organism using Vinogradskii's silica gel plates to observe the "very characteristic growth of the organism." Ibid., 305. Romell

At Cornell, Romell had frequent opportunities to discuss Vinogradskii's direct method with soil scientists. For example, in his numerous meetings with Selman Waksman and H. J. Conn, the conversation often turned to the usefulness of the direct method for soil science and the history of Vinogradskii's work.[96] Clearly seeking Vinogradskii's approval, Romell recounted one such meeting in a letter to him in 1935.[97] At the symposium on agricultural microbiology convened in Amsterdam in 1935, a group of internationally renowned microbiologists debated the significance of correlating the role bacteria play in soil processes with the amount (or proportion) of bacteria in the soil.[98] In a conversation that "visibly lasted too long" for their audience, Romell, Waksman (President of the meeting), Kluyver, Rossi (of the Kholodnyi-Rossi method), Fernand Chodat (a Swiss bacteriologist), and Augusto Bonnazi (a Venezuelan microbiologist) debated the relative worth of pure cultures and Vinogradskii's direct method. Kluyver placed the direct method in a somewhat supplementary position to pure cultures, and Chodat even more so—inspiring Romell to speak at length on the practical applications of the direct method, the results Vinogradskii had achieved by using it in his *Azotobacter* investigations, and its efficacy for exploring the "ecological character of soil microbiology."[99] When the dust settled, Romell's speech had moved Bonnazi to characterize the direct method as preferable to the "fixed methods"—making one more convert for his Russian mentor.

---

[96] See the correspondence between Romell and Vinogradskii.

[97] Letter from Romell to Vinogradskii, September 13, 1935 from Noppikoski, Alvho, Sweden. Winogradsky Papers, *Archives de l'Institut Pasteur*, Box WIN 2, L. G. Romell Folder.

[98] Romell informed Vinogradskii that they discussed the *Azotobacter* work Vinogradskii had conducted with Ziemiecka. Ibid., 1.

[99] Ibid., 2.

Chapter 10
Vinogradskii's Reception in Russian and Soviet Microbiology

Vinogradskii was a Russian scientist. The simple restatement of this fact, however, only begins to explain the wide recognition he enjoyed during the Soviet period—and still enjoys today. In the last decades of Imperial Russia, he had gained a reputation as a leading figure in world science through his novel scientific discoveries, and by contributing significantly to the institutional and disciplinary development of Russian science. During his second, French career, he never returned to Russia, yet Soviet scientists continued to cite his publications. Even during the Soviet period, when émigrés like Vinogradskii could be dangerous contacts, a significant number of Soviet scientists maintained correspondence with him and visited his laboratory in Brie-Comte-Robert.

As elsewhere in the world, Vinogradskii initially was known in the Soviet Union primarily for his discovery of autotrophism and chemosynthesis. Soviet agricultural chemists, biogeochemists, soil scientists, and ecological microbiologists cited Vinogradskii alongside other foundational figures in the history of microbiology, such as Martianus Beijerinck, Louis Pasteur, and Robert Koch. Being a world-renowned scientist, the Soviet government may have found him to be a useful propaganda tool.[1] Familiarity with Vinogradskii's earlier research on autotrophic microorganisms and nutrient cycles in nature often led Soviet scientists to develop an interest in his ecological methods. In this way, reports of Vinogradskii's new ecological microbiology quickly spread through the Soviet scientific community. Just as in the West, Soviet scientists working in a wide range of disciplines, appropriated Vinogradskii's vision and methods, and his still considerable authority, to promote their own projects.

[1] It is somewhat telling to compare three Russian scientists of the same generation—Vinogradskii, Vernadsky and Ivan Pavlov. Vernadsky returned to Russia in 1926, and became a great organizer of Soviet science in the areas of biogeochemistry and related disciplines. Had he decided to remain in Paris in 1926, he may have experienced Vinogradskii's fate—a somewhat accepted émigré scientist. Ivan Pavlov was also of the same generation and, having weathered the changes to a Soviet state and scientific system within the country, he experienced a quite different situation from both Vernadsky and Vinogradskii.

L. Ackert, *Sergei Vinogradskii and the Cycle of Life: From the Thermodynamics of Life to Ecological Microbiology, 1850-1950*, Archimedes 34,
DOI 10.1007/978-94-007-5198-9_10,

Vinogradskii's First Student, V. L. Omelianskii: Microbes as "Living Reactives"

Among Vinogradskii's few Russian students at the IEM, one stood out. Vasilii Omelianskii proved to be an adept Vinogradski-ite. He not only adopted and developed Vinogradskii's methods in his own research, but he also passed this training on to new generations of Russian microbiologists, physicians, and soil scientists.[2] Thus educated in Vinogradskii's ecological approach, they translated the cycle of life vision into new research directions in ecological microbiology, medical bacteriology, and soil science.

Vinogradskii took a risk when he hired Omelianskii to be his first laboratory assistant in 1893; Omelianskii had little or no experience in microbiology. After earning a BA in chemistry from St. Petersburg University in 1890 under N. A. Menschutkin, he decided not to pursue a doctorate and, instead, went to work in a metallurgy factory. [3] While working in industry, Omelianskii kept alive his interest in chemistry and set upon an independent study of its application in soil science.[4] Omelianskii had

---

[2] The list of praktikanty and visitors to Department of General Microbiology at the IIEM was not as long as for other departments. At this time, Pavlov, for example, could count on the assistance of dozens of praktikanty. Yet the several that worked there went on to prominent positions in Russian science. Omelianskii, for example, directed the work of M. N. Rubel' on the question of nitrification in biological filters (1913), E. I. Nikolaevna in her investigations of actinomycetes, S. N. Bukhbinder on pigment formation in colorless bacteria, N. M. Perepelitsina on the use of triptophan reactions in bacteriology, I. A. Makrinov on the aerobic ferments of nicotinic materials, O. M. Bogoliubov on the bacteriology of lactobacillus, L. D. Shturm on sapropel (an aquatic sludge rich in organic matter), Germanov on methods for investigating the soil microflora, and V. P. Neelov on the chemistry of the nitrification process (1922). On the history of the Department of General Microbiology at the IEM, see: B. L. Isachenko, "Otdel Obshchei Mikrobiologii," *Material k Istorii V.I.E.M.* (Moskva, 1941), 87–103; See 96.

[3] Z. G. Razumovskaia, "Zhizn' i nauchnaia deiatel'nost' Vasiliia Leonaidovicha Omelianskogo," *Russkie Mikrobiologi S. N. Vinogradskii i V. L. Omelianskii* (Moskva: Izadatel'stvo Ministerstva Sel'skogo Xoziaistva SSSR, 1960), 17–25, see 17. Omelianskii studied with N. A. Menschutkin—the influential Russian chemist—and conducted a laboratory investigation on "the question of the influence of the dilution on the speed of chemical reactions." Ibid.

[4] In 1891, Omelianskii took copious notes on two important Russian agrochemical works: Grando's, "Novaia mineral'no-gumusovaia teoria pitaniia rastenii (A new mineral-humus theory for plant nutrition)," published in *Sel'skoe Xosiastvo i Lesovodstvo (Agriculture and Forestry)*, (1872, 1873); and G. G. Gustavson's, "20 Lektsii Agronomicheskoi Khimii (20 Lectures on Agricultural Chemistry)." Gustavson worked at the Moscow Agricultural Institute at the end of the nineteenth century, where he developed chemical investigations related to V. V. Dokuchaev's "concept of the soil as a natural body." For a recent history of Russian soil science see, I. V. Ivanov, *Istoriia Otechestvennogo Pochvovedeniia: Kniga Pervaia, 1870–1947* (Moskva: Nauka, 2003); on Gustavson at the Moscow Agricultural Institute, see 137. In addition, Omelianskii had read the works of the founding fathers of Russian soil science including numerous articles and books by V. V. Dokuchaev, such as his "Kratkaia programma dlia izsledovaniia pochva (A short program for the investigation of the soil), in which Dokuchaev proposed methods for investigating soils as organisms; and P. A. Kostychev's, "Pochvy chernosemnoi oblasti Rossii, ix proiskhozhdenie, sostav i svoistva (The soils of Russia's black earth region, their origin, make-up, and characteristics)," (St. Petersburg, 1886). See Omelianskii's notebooks, St. Peterburgskii Archiv Russkoi Akademii Nauk, fond 892, opis' 1, delo 13, listy 1–62 oborot.

published only a single article based on his undergraduate investigation of "the influence of dilution on the speed of chemical reactions."[5] Omelianskii had continued his study of chemistry and its applications to soil science. Menschutkin noticed, however, that Omelianskii possessed considerable talent in chemistry, enough to recommend him to be Vinogradskii's assistant. Vinogradskii "had grown accustomed to respecting [Menschutkin's advice] since his own student days" and his old chemistry professor's sanction in this matter was more than adequate.[6]

Vinogradskii assigned his new assistant practical work, believing it to be the fastest rout to technical proficiency. When he arrived in Vinogradskii's laboratory in 1893, Omelianskii possessed an extensive knowledge of agricultural chemistry, but a limited understanding of the biology of the soil.[7] He recalled that he "quickly and imperceptibly mastered the bacteriological techniques" and that his "knowledge of chemistry provided a good basis for investigating microbial physiology."[8] It seemed to him that during this training, "a 'microbiological wisdom (*mudrost'*)' arose on its own (*prishla sama soboi*)".[9] It is clear, however, that he did not assimilate his wisdom from some mysterious atmosphere—Omelianskii had adopted entirely Vinogradskii's approach.

Omelianskii's participation in Vinogradskii's ongoing investigations gave him not only the opportunity to practice microbiological techniques, but also framed his future scientific interests. He joined the laboratory when Vinogradskii was investigating the assimilation of gaseous nitrogen by microbes. It was during this research, that Vinogradskii developed the elective culture method and isolated the first freely living nitrogen-fixing organism, *Clostridium pasteuranium*. The questions that drove this research and the methods by which they were explored were Vinogradskii's. Omelianskii, however, conducted the bulk of the analyses.[10] In the process, he learned first-hand Vinogradskii's method for developing laboratory techniques to explore the role of microbes in the cycle of life. By the end of the century, Omelianskii was publishing independent articles, yet these still reflected the laboratory's direction. He collaborated with Vinogradskii to promote the elective culture method for investigating nitrification and isolating nitrifying microbes from the soil.[11]

[5] Ibid.

[6] S. N. Vinogradskii, "Pamiati V. L. Omelianskogo: Lichnye Vospominaniia," *Rasskazy o Velikom Bakteriologe S. N. Vinogradskom* (St. Petersburg: OOO "Izdatel'stvo "Rostok"", 2002), IU. A. Mazing, T. V. Andriushkevich, IU. P. Golikov eds.; 306–309; see 306 for quote.

[7] The early soil science literature largely ignored the biological components of the soil and only very rarely discussed any investigations into it. Most of these works cited the work of Pasteur and Charles Darwin on the formation of soil.

[8] Ibid.

[9] Ibid., 18.

[10] S. N. Vinogradskii, "Ob usvoenii svobodnogo azota atmosfery mikrobami," *Arkhiv Biologicheskikh Nauk*, 1895, Vol. 3. No. 4, 293–351.

[11] V. Omelianski, "Ueber die Isolierung der Nitrifikationsmikroben aus dem Erdboden," *Centralblatt fur Bakteriologie, Parasitenkunde und Infektionskrankheiten*, 2nd Abteilung, Bd. V., No. 15, 31 Juli 1899, 537–549; Idem., "Ueber die Nitrifikation des organischen Stickstoffes," Bd. V, No. 13,

Even after 1898, when Vinogradskii spent increasingly less time in the laboratory for health reasons, Omelianskii continued to investigate the lines of research established by his mentor.[12] In the twentieth century, Omelianskii applied Vinogradskii's elective culture method to still unexplored areas of microbiological research. At the heart of Omelianskii's work was Vinogradskii's notion that a microbe's physiological characteristics determined (or reflected) its role in the transformation of matter in nature. Omelianskii synthesized his mentor's cycle of life perspective with his own chemistry training, leading him to envision the role of microbes in nature as "living reactives."[13]

Through his institutional activities, mentoring, and publications, Omelianskii continued Vinogradskii's efforts to expand microbiology's place in Russian science. In the first decades of the nineteenth century, Vinogradskii moved increasingly into retirement leaving, Omelianskii to assume his duties.[14] After 1906, for example, Omelianskii served temporarily as director of the IEM's Department of General Microbiology (a position that became permanent upon Vinogradskii's resignation in 1912). During this period, Omelianskii "preserved a great scientific heritage;" through his numerous investigations and "outstanding" teaching, he exerted "a great influence on the development of microbiological sciences in [Russia]."[15] In 1906, he accepted an invitation to teach microbiology at the newly organized Lokhvitskoi-Skalon advanced natural history courses for women.[16] In 1909, he published the first Russian textbook on general microbiology, *Osnovy Mikrobiologii* (*The Fundamentals of Microbiology*), which went through nine (including several posthumous) editions. In all of this work, Omelianskii discussed extensively Vinogradskii's role in the history of microbiology.

After 1912, events separated Omelianskii and Vinogradskii. While on a trip to Dresden, Omelianskii learned through a colleague that Vinogradskii had officially resigned from the IEM. Although Omelianskii was expecting this "formality," since

15 Juni 1899, 473–490; S. Winogradsky und V. Omelianski, "Ueber den Einfluss der organischen Substanzen auf die Arbeit der nitrifizierenden Mikrobien," *Centralblatt für Bakteriologie, Parasitenkunde und Infektionskrankheiten*, Bd. V, No. 10, 329–343; Idem., No. 11, 377–387; Idem., 429–440.

[12] In 1898–99, Vinogradskii developed kidney disease (nephritis) after a severe case of the flu. He began to avoid St. Petersburg's climate—which he had always found to be harsh—for the more comfortable climate of his Ukrainian estate. Selman Waksman, "Sergei Nikolaevich Winogradsky, September 1, 1856-August 31, 1946: The Story of a Great Bacteriologist," *Soil Science*, Vol. 62, No. 3, September, 1946, 196–226; See esp. 209–210.

[13] V. L. Omelianskii, *Osnovy Mikrobiologii* (S-Peterburg, Tipografia Imperatorskoi Akademii Nauk, 1909), see esp. chapter 25, "Krugovorot' veshchestv v prirode," 188–195.

[14] For information on the frequency of Vinogradskii's absences from the IEM, see the Departmental reports for this period: Leningradskii Gosudarstvennyi Istoricheskii Arkhiv, St. Peterburg, fond 2282 "Imperaterskoi Institut Eksperimental'noi Meditsiny," opis' 1 "Delovye Otchety," in several dela.

[15] Z. G. Razumovskaia, "Zhizn' i nauchnaia deiatel'nost' Vasiliia Leonaidovicha Omelianskogo," 18, 24. In 1909, for example, Omelianskii taught the Vinogradskii approach to Nikolai Kholodnyi.

[16] Ibid., 24.

Vinogradskii had actually left the Institute years earlier, the news "provoked in [him] very complex feelings, primarily of a melancholy shade. It was as though something very dear snapped off and ended forever—that was the thread that bound me with you into one spiritual whole, into a single common scientific organism."[17] He would, Omelianskii pledged, "remain a student, the natural prolongation of [Vinogradskii's] spiritual heritage."[18] Between 1912 and 1922, when Vinogradskii was living the "life of a latent scientist," Omelianskii continued to remind his students and colleagues of the value of his mentor's work.[19]

In the 1920s, Soviet officials allowed Omelianskii to publish the research of his now expatriated colleague.[20] Throughout this period, Omelianskii not only retained his directorship at the IEM department of general microbiology, he also increasingly earned the esteem of the Russian scientific community. He may have earned his substantial institutional power by demonstrating how his work—and that of Vinogradskii's—had great value for agriculture and medicine. V. I. Lenin, for example, had become interested in the practical applications of microbiology, and communicated with Omelianskii about his research on the role of microbes in increasing soil fertility.[21] Omelianskii had discussed this topic in 1925 in a survey of the history of Russian microbiology.[22] Here he recalled not only Vinogradskii's classic work on iron, sulphur, and nitrogen bacteria, but also his new work on the bacteriological strength, or fertility, of the soil.[23]

Finding his promise "to prolong Vinogradskii's heritage" somewhat premature—the Master had returned to science and was causing a noisy storm from Brie-Comte-Robert—Omelianskii began instead to promote Vinogradskii's new ideas. In 1926, Omelianskii outlined Vinogradskii's direct method for the readership of the State Institute for Experimental Agronomy.[24] Here he endorsed the direct method as one of "the newest trends [for] investigating the microbial strength of the soil"—better

---

[17] Letter from Omelianskii to Vinogradskii, 21 September 1911; Arkhiv Russiiskoi Akademii Nauk, Peterburgskoi Filial, fond 1601, opis' 1, delo 158, listy 13–14 oborot. See List 13 oborot for quote.

[18] Ibid.

[19] In a letter to Omelianskii in 1922, Vinogradskii characterized his decade outside science as "la vie latente d'un savant." V. L. Omelianskii, "Sergei Nikolaevich Vinogradskii: Po povodu 70-letiia so dnia rozhdeniia," *Arkhiv Biologicheskikh Nauk*, 1927, Tom. XXVII, Nos. 1–3, 11–36; See 24 for quote.

[20] In 1916 he was selected corresponding member of the Russian Academy of Sciences, in 1917 St. Petersburg University awarded him an honorary doctorate in botany, and that same year he was elected full member of the RSFSR Academy of Sciences. Z. G. Razumovskaia, "Zhizn' i nauchnaia deiatel'nost' Vasiliia Leonaidovicha Omelianskogo," 18.

[21] B. L. Isachenko, "Otdel Obshchei Mikrobiologii," *Material k Istorii V.I.E.M.* (Moskva, 1941), 97.

[22] V. L. Omelianskii, "Puti Razvitiia Mikrobiologii v Rossii," *Izbrannye Trudy*," (Moskva: Izdatel'stvo Akademii Nauk SSSR, 1953), Tom II, 38–46.

[23] Ibid., 43.

[24] V. L. Omelianskii, "Noveishie Techeniia v Oblasti Issledovaniia Mikrobnykh Sil Pochvy," *Izbrannye Trudy*," Tom II, 47–51; See 50–51 on the direct method.

than the methods developed by "the Americans" such as Lipman, Waksman, and Conn. For Omelianskii, "Vinogradskii, with his characteristic straightforwardness, decided to break sharply with the fixed pattern (*shablon*) and to begin the investigation anew, selecting as his objective the soil itself with all of its diverse populations."[25] Vinogradskii had conducted his work, Omelianskii continued, with "heated excitement and bravery—the bravery of a reformer."[26] "Henceforth, Omelianskii declared, Vinogradskii's "direct method" would be generally accepted," and would be applied alongside previous methods for analyzing the microbiology of the soil.[27]

Omelianskii and Vinogradskii formed a strong bond in the 1890s. Their relationship made Omelianskii the best candidate to write a biographical article celebrating Vinogradskii's seventieth birthday.[28] The following year, Vinogradskii had the gloomy honor of returning the favor—Omelianskii died in April of 1928 after a serious illness.[29] Twenty-five years later, Omelianskii's collected works appeared 1 year after Vinogradskii's, and in 1960 the Soviet Ministry of Agriculture celebrated their contributions to Russian science with the volume, *The Russian Microbiologists, S. N. Vinogradskii and V. L. Omelianskii.*[30]

Anecdotal evidence that Vinogradskii's Russian audience appreciated his approach appears in a letter informing him of his selection as an honorary member of the Microbiological Society. In 1910, on the occasion of the society's fiftieth meeting, Vinogradskii's students Omelianskii and D. Zabolotnyi—both prominent scientists in their respective fields—informed their teacher that:

> Your closest students, not losing hope for collaborative work in the future, with a feeling of lively happiness are sending to you the corresponding diploma, which grants you, along with I. I. Metchnikov, E. Roux, A. La'veran, R. Koch, E. Behring and D. Ehrlich, an honorary role in the *cycle of life* of the Microbiological Society.[31]

---

[25] Ibid., 50.

[26] Ibid.

[27] Ibid.

[28] Although relatively short, this work greatly influenced all the subsequent biographies on Vinogradskii—including this one. V. L. Omelianskii, "Sergei Nikolaevich Vinogradskii: Po povodu 70-letiia so dnia rozhdeniia," *Arkhiv Biologicheskikh Nauk*, 1927, Tom. XXVII, Nos. 1–3, 11–36.

[29] S. N. Vinogradskii, "Pamiati V. L. Omelianskogo: Lichnye Vospominaniia," in Iu. A. Mazing, T. V. Andriushkevich, Iu. P. Golikov eds., *Rasskazy o Velikom Bakteriologe S. N. Vinogradskom* (St. Petersburg: Rostok, 2002), 306–309.

[30] S. N. Vinogradskii, *Mikrobiologiia Pochvy: Problemy i Metody, Piat'desiat Let Issledovanii* (Moskva: Izdatel'stvo Akademii Nauk, 1952), and V. L. Omelianskii, "Noveishie Techeniia v Oblasti Issledovaniia Mikrobnykh Sil Pochvy," *Izbrannye Trudy*," (Moskva: Izdatel'stvo Akademii Nauk SSSR, 1953).

[31] In a letter of March 8, 1910, Arkhiv Rossiiskoi Akademii Nauk, Petersburgskoi Filial, fond 1601, delo 140, listy 1–2. My italics. The letter was signed also by several other important scientific workers in Russian microbiology and bacteriology.

Nikolai Kholodnyi: Iron Bacteria Research and Vinogradskii's Direct Method

While directing the Laboratory of General Microbiology at the IEM, Omelianskii taught numerous praktikanty and students. Nikolai Kholodnyi, for example, at the beginning of his training in plant physiology visited Omelianskii specifically to study microbiological methodology. Kholodnyi expressed an early interest in the natural sciences, spending his childhood in Voronezh and his gymnasium education in Novocherkassk collecting insects and studying ornithology.[32] In 1900, he entered Kiev University, where he would spend much of the next 40 years. Initially studying ornithology, under the guidance of the plant physiologist Konstantin Purievich, Kholodnyi took up the study of the geotropic sensitivity of roots.[33] Upon graduation in 1907, he worked as an assistant in Purievich's Department of Plant Physiology and Anatomy. Participating in Purievich's research on respiration in fungi and particularly in the filamental microscopic fungi *Aspergillus niger* and the energetics of photosynthesis, Kholodnyi would have encountered Vinogradskii's research.[34]

Like many Russian students, Kholodnyi spent a year studying in Europe, after which he returned to Kiev University. His arrival in 1908 coincided with appearance of the first microbiology course in the university's curriculum. He took an immediate interest in the subject and "actively participated in the organization and implementation of practical lessons (*zaniatiia*)."[35] Over the next few years, he earned his master's degree (1912) and Purievich invited him to teach the microbiology course. In order to improve his knowledge of microbiological methods of investigation he studied with Omelianskii at the IEM.[36]

It was when Kholodnyi turned his full attention to microbiology that he came to appreciate the significance of Vinogradskii's discoveries and approach. In 1912—when working with Omelianskii—Kholodnyi discovered the existence of thermophilic denitrifying bacteria.[37] In 1915, Kholodnyi began to study iron bacteria—a subject Vinogradskii pioneered in 1888 and that Omelianskii was pursuing in the 1910s. In his iron bacteria investigations, Kholodnyi supported Vinogradskii's opinion that if they always appeared in oligotrophic environments—that is, ponds, lakes, or bogs, poor in plant material and rich in dissolved oxygen—they must be autotrophs.[38] Kholodnyi strongly defended Vinogradskii's monomorphist position

[32] On Kholodnyi, see A. A. Imshenetskii, "N. G. Kholodnyi i ego Mikrobiologicheskie Issledovaniia," forward to N. G. Kholodnyi, *Zhelezobakterii* (Moskva: Izdatel'stvo Akademii Nauk SSSR, 1953), 3–16.

[33] Ibid., 5. On Purievich see "Konstantin Adrianovich Purievich, (1866–1916)," *Biologi: Biograficheskii Spravochnik* (Kiev: Naukova Dumka, 1984), 519.

[34] Ibid.

[35] Imshenetskii, "N. G. Kholodnyi i ego Mikrobiologicheskie Issledovaniia," 5.

[36] Ibid.

[37] Ibid., 6.

[38] Ibid., 8.

that, morphologically and physiologically, these autotrophic bacteria were defined by their ecological relationships—their role in the environment—and could not exhibit a wide variation in characteristics. Later ecological microbiologists, such as Alexander Imshenetskii, found the arguments Kholodnyi made in support of Vinogradskii's views to be "momentous (*veskie*) and completely justified from an ecologically-physiological point of view.[39]

Russia's Civil War would lead Kholodnyi to develop close, and unexpected ties with an influential group of scientists, and to expand his iron bacteria research. Having just completed his master's degree in 1919, Kholodnyi found a position as director of the Dnepr Biological Station in Starosele, Ukraine. Located 18 km outside of Kiev, this scientific institution provided a relatively secure refuge from the "Red Terror" that had descended upon the area.[40] The threat was genuine. When the Bolsheviks were in control of the city in the spring and summer of 1919, they began a program of taking members of the middle class and the old regime hostage and even executed hundreds of them.[41] Vladimir Vernadsky, a geochemist well-known today as creator of the concept of the biosphere, president and co-founder of the Ukrainian Academy of Sciences, and a political activist, began to fear for his and his family's safety and moved to the station.[42] According to his research assistant, Theodosius Dobzhansky—who became an important geneticist in the 1930s—considered Vernadsky one of the foremost scientific personalities in the Kiev area.[43]

While at the station, Vernadsky and Kholodnyi started a "strong intellectual friendship" that would remain close for 40 years.[44] At the time, Vernadsky was investigating the geochemical properties of nature, and especially the role of "living matter." He conducted a series of experiments on the chemical composition of different types of organisms with the objective of categorizing all life chemically. Additionally, he developed a theoretical foundation for the new science of biogeochemistry, including the concepts of the speed of life, the prevalence of life, the

---

[39] Here Imshenetskii evaluated Kholodnyi's research as a continuation and validation of Vinogradskii's theory of chemosynthesis. Moreover, writing in 1953, Imshenetskii still considered chemosynthesis to be a viable line of investigation with unanswered questions. For example, it still remained to determine "the mechanism of oxidation of inorganic compounds, in particular, the ferment systems . . . the vague fact of the discovery of such classic autotrophs as *Nitrosomonas* in the silt on the bottom of the ocean, or in environments very rich in organic matter . . . and the physiology of pure iron bacteria cultures." Ibid., 9.

[40] For a description of the "Red Terror" in Kiev and how some scientists escaped it see: Theodosius Dobzhansky, *The Reminiscences of Theodosius Dobzhansky*, (Oral History Research Office, Columbia University, 1962), 65–72.

[41] On Vernadsky's experiences during this period see Kendall E. Bailes, *Science and Russian Culture in an Age of Revolutions: V. I. Vernadsky and His Scientific School, 1863–1945* (Bloomington and Indianapolis: Indiana University Press, 1990), 141–148.

[42] Ibid., 144.

[43] Theodosius Dobzhansky, *The Reminiscences of Theodosius Dobzhansky*, 65–72. Dobzhansky, who worked as Vernadsky's research assistant, made weekly trips carrying mail and other supplies to the station.

[44] Bailes, *Science and Russian Culture in an Age of Revolutions*, 145–146.

pressure of life, and its adaptability.[45] Vernadsky had taken particular interest in Vinogradskii's autotrophic work—it provided a method for understanding and investigating how living organisms contributed to the concentration of chemical elements in nature. Vinogradskii's discovery of iron bacteria, for example, became an important aspect of Vernadsky's biogeochemistry. After learning that Kholodnyi had studied iron bacteria in Vinogradskii's laboratory with Omelianskii, Vernadsky urged Kholodnyi to pursue that topic.[46]

Kholodnyi's iron bacteria research in the 1920s led him again to Vinogradskii and his direct method. In 1922, Vinogradskii published his first research in a decade. For the topic of this work, Vinogradskii went back to his earliest investigations—on iron bacteria. Reacquainted with this classic Russian scientist, Kholodnyi followed Vinogradskii's publications and like many others, thought the direct method held great promise for reforming soil science.[47]

The Role of Microbes in Russian and Soviet Soil Science

In the second half of the nineteenth century, Russian scientists revolutionized the study of the soil. As we have seen above, one of soil science's founders, Vasilii Dokuchaev, was the first to propose considering each kind of soil as an organism comprised of specific defining characteristics. He built the first coherent system for categorizing, studying, and managing these Russia's soils. Within this system he recognized the importance of understanding the role of microorganisms in soil formation. His colleague and another founder of Russian soil science, Pavel Kostychev, made microbiology an essential part of his science in its earliest days. Much like Vinogradskii, he studied with Louis Pasteur in the 1860s–1870s and applied his training to study how microbes decompose cellulose in the soil. An integral part of early soil science, the study of cellulose decomposition eventually became an essential tool in agriculture, eventually providing a method for determining the soil fertility.[48]

---

[45] Ibid., 145.

[46] Kholodnyi dedicated the reprinting of his monograph, *Zhelezobakterii* (1952 to his close friend Vladimir Vernadsky, who had first stressed the importance of the topic to him. Kholodnyi first published *Zhelezobakterii* in 1926.

[47] N. G. Kholodnyi, "Kak Nabliudat' Zhizn' Mikroorganismov Pochvy," in *Sredi Prirodi i v Laboratorii* (Moskva: Izdatel'stvo Moskovskogo Obshchestva Ispytatelei Prirody, 1949), 101–121.

[48] For some examples of this see S. P. Kostychev, Kovda, and Krasil'nikov. In 1927, while on an official trip, Sergei Kostychev visited Vinogradskii in Brie-Compte-Robert. The primary purpose of Kostychev's trip was to obtain scientific instruments ordered from companies in Berlin. There was also a scientific dimension to his expedition—after visiting Bertrand and Fernbach at the Pasteur Institute in Paris to discuss fermentation, Kostychev made his way out to Vinogradskii's laboratory where he discussed soil microbiology. Kostychev reported to superiors that his conversations with Vinogradskii "were related to methods for the quantitative calculation of soil fertility factors, developed on the fundamentals Vinogradskii had established for soil microbiology." (For a description of Kostychev's activities on this trip see, "Otchet o deiatel'nosti S. P. Kostycheva i komandirovka za

Vinogradskii took up this line of research in at the IEM in the1890s, assigning his praktikants to investigate the role of microbes in cellulose decomposition.[49] Vinogradskii also returned to this subject when at Brie-Comte-Robert.

Dokuchaev taught his holistic soil science to a generation of Russian scientists—and through them to many in the world scientific community. Vernadsky, one of his best known students, developed Dokuchaev's system into an expansive program encompassing the entire realm of the Earth. Vernadsky introduced Vinogradskii's work to geologists and geographers. The founder of geochemistry and its related discipline of biogeochemistry, Vernadsky is best well-known in the West for his concept of the "biosphere."[50] The biosphere for Vernadsky was the area on the Earth where life existed or could exist.[51] The most novel aspect of his biosphere concept was his multidisciplinary program for studying life as an historical, geological, biological, and chemical phenomenon. Vernadsky first made public his biosphere concept in a series of lectures at Kiev University (1918–1919), the Academy of Sciences in St. Petersburg (1921), at the Petrograd Agricultural Institute (1921–1922), and finally, while on temporary assignment to the Curie laboratory at the Sorbonne in Paris (1923–1924). He published his lectures in 1924 as "*La Géochimie*" and in 1926 as "*La Biosfera*."

Making the very difficult decision not to emigrate, Vernadsky returned to the Soviet Union 1926. There he became one of the most powerful scientific organizers in the Soviet Union through the early 1950s. Through his institutions Vernadsky exercised great influence on Soviet science—an influence that simultaneously promoted Vinogradskii's discoveries and research.

Vernadsky incorporated Vinogradskii's concept of autotrophism into his understanding of the geochemical evolution of the environment.[52] Autotrophism—the ability of organisms to "live independent of the energy of light, because they get the energy for their vital processes from minerals"—was a crucial component of Vernadsky's concept of the biosphere concept.[53] In particular, Vinogradskii's aided Vernadsky in portraying the "geochemical history of living matter," by providing him a mechanism—autotrophic bacteria—for understanding how "living matter transports chemical elements through the biosphere."[54] For Vernadsky, Vinogradskii's discovery of autotrophism and his array of laboratory techniques made it possible to

---

granitsu v 1927 g.," fond 159, opis 1, dela 70, listy 5–6; See list 5 for quote.) Kostychev had high hopes for Vinogradskii's research and expected "several new methods (*priemy*) would be introduced into this method (*metodika*) on the basis of the results received." Ibid., 5.

[49] For a discussion of Vinogradskii's cellulose decomposition research at the IEM, see Chap. 3.

[50] He first published his views of the biosphere in Vladimir Vernadsky

[51] Vernadsky was not the first to use the term biosphere. Lamarck had defined it as any place life existed and the geologist Eduard Suess defined it as the geological strata produced by living organisms.

[52] Vladimir I. Vernadsky, *Trudy po Geokhimii*, ed. A. A. Iaroshevskii (Moskva: Nauka, 1994), 98–101.

[53] Vladimir I. Vernadsky, *Zhivoe Vechestvo* (Moskva: Gosizdat, 1930), 88.

[54] Ibid., 89.

investigate the migration of chemical elements throughout the biosphere. Synthesizing Vinogradskii's approach with his own mathematical and chemical techniques, Vernadsky could now investigate how autotrophic organisms concentrated various elements in nature. Moreover, it provided him another way for understanding the dynamics between the inorganic and organic realms of the biosphere.

Ecological Microbiology in the Soviet Union

After the end of Stalin's reign, scientists felt freer to recognize émigré Russian scientists, including Vinogradskii. A testament to his Russian and Soviet legacy came in the form of collaborative publications celebrating his scientific contributions during the 1960s and later. For example, in 1960, the Leningrad Department of the Society of Microbiologists, Epidemiologists and 'Infectionists' published a collection of articles recognizing his role as a founder of general and soil microbiology in both Russian and international science.[55] It comes as no surprise that these authors were well versed in Vinogradskii's ecological perspective and methods. Most of their information came from the 1952 Russian translation of Vinogradskii's French *Microbiologie du Sol* that had appeared in 1949. In much the same language used by Vinogradskii, V. N. Bylinkina describe his emergence from retirement in 1922, his critique of the standard methods in microbiology, and his development of an ecological method. Where Vinogradskii left off, however, Bylinkina continued.

She described how "Vinogradskii's views on the necessity of an ecological approach in the study of soil microflora found wide recognition (*priznanie*) and were reflected in the works of a series of investigators."[56] G. L. Seliber, the editor of the volume, emphasized that Vinogradskii's methods to "study microorganisms in the natural environment in which they grow," greatly affected the growth of agricultural microbiology, and especially his own development of an ecological approach to that science.[57] In addition, H. M. Lazarev and his colleagues built upon Vinogradskii's work when formulating their conception of "the soil bioorgano-mineral complex as a complex system of organism-environment and on the ecological grouping of soil microflora."[58] Bylinkina similarly attributed E. N. Mishchustin's view of soil microorganism ecology to Vinogradskii's notion of "the geological distribution in various soils."[59] To understand Vinogradskii's legacy we can make a detailed analysis of

---

[55] G. L. Seliber, ed., *Russkie Mikrobiologi S. N. Vinogradskii i V. L. Omelianskii* (Moskva: Izadatel'stvo Ministerstva Sel'skogo Xoziaistva SSSR, 1960). The authors of these collected articles attempted to draw connections between the work of Vinogradskii and Omelianskii and contemporary developments in their own disciplines.

[56] V. N. Bylinkina, "Metody pochvenno-mikrobiologicheskikh issledovanii v rabotakh S. N. Vinogradskogo i ikh dal'neishee razvitie," in Idem., 31–37; see 37.

[57] Ibid.

[58] Ibid.

[59] Ibid.

whether or how these scientists actually drew upon Vinogradskii's work. Just as important, however, is that: "Vinogradskii's view that soil microbiology would progress only on the basis of ecological methods of investigation" became justified and actualized in their own contemporary research.[60]

Russian and Soviet soil science could hardly have avoided Vinogradskii's work. V. P. Vushinskii attributed V. R. Vil'iams's great role the development of agricultural and biological sciences to his familiarity with the well-known work of Pasteur, and Vinogradskii and Omelianskii.[61] Seliber's volume in 1960 shows that Russian, Soviet, and Western soil scientists from a number of sub-disciplines engaged and developed Vinogradskii's research.[62]

In the 1970s, S. N. Kuznetsov—director of the Soviet Academy of Sciences' Institute of Microbiology in Leningrad—synthesized Vernadsky's biosphere concept with Vinogradskii's ecological microbiology. Ecological microbiology, he wrote, was based on Vinogradskii's that microorganisms are tightly connected with the environment and that they should be studied as part of the dynamic biological processes in nature. At the basis of contemporary ecological microbiology, lay Vinogradskii's principle that the investigation of real processes carried out by microbes in nature, should be founded not only on the study of the behavior of individual microorganisms, but also on the behavior of microbial associations as a whole.

In 1949, Vinogradskii made one final impact on soil microbiology and microbial ecology. In that year, with the help of Selman Waksman, he published an 800-page compendium of his life's work: the French monograph *Soil Microbiology: Problems and Methods, Sixty Years of Research*. This was no mere republication of his most important papers and speeches—Vinogradskii edited this literature, rewriting himself into the history of ecology. New readers on an international scale, captured in part by Waksman's 1953 biography of Vinogradskii, and by the Russian and Polish translations of his massive compendium, were attracted to his strong, authoritative voice that echoed their own interests in ecology.

[60] Ibid.

[61] V. P. Vushinkii, "Rol' V. R. Vil'iamsa v razvitii sel'skokhoziastvennykh i biologicheskikh nauk," in *V. R. Viliams, Sobrannye Sochinenii, Tom 1, Raboty po Pochvovedeniiu (1888–1902)* (Moskva: Sel'xozgiz, 1948), 7–38, esp. 17. For Vushinskii, the most significant aspect of their work was "the role of microorganisms in the decomposition of animal remains, which are located in and on the soil." Ibid.

[62] G. L. Seliber, ed., *Russkie Mikrobiologi S. N. Vinogradskii i V. L. Omelianskii.*

Chapter 11
Conclusions

In this work, I have woven together historical threads from Vinogradskii's life and scientific work to describe a new dimension of the history of ecology. Lying along an obscure trajectory in that history, Vinogradskii's story accentuates the role—not of natural historians, Darwinists, and plant communities—but rather of experimentalists (who often fused their laboratory investigations with field observations), holists, and soil microbes. As this cast of characters suggests, incorporating this dimension of the history of ecology into the larger disciplinary story requires that we include a neglected set of scientists who perceived themselves as pursuing an experimental investigation of energy, matter, and life. By recognizing the influence of Pasteur, Famintsyn, Dokuchaev, Vinogradskii, and other such late-nineteenth century microbiologists, plant physiologists, and soil scientists, historians of science will derive a fuller and ultimately more satisfactory account of the historical development of ecology.

The essence of this story is how scientists negotiated the shifting relationship between natural history and laboratory research. Vinogradskii's career exemplifies how scientists could balance their commitments to romantic ideals associated with natural history, on the one hand, with the escalating interest in another kind of knowledge—the ideal of experiment, on the other. In the first half of the nineteenth century, efforts to categorize nature slowly yielded to increasingly dynamic natural historical systems including Humboldt's phytogeography, Lyell's historical geology, and Darwin's theory of evolution. The second half of the century was characterized by an increased reliance on the laboratory, which—with its focus on experiment and the physical and chemical investigation of organic and inorganic bodies—threatened natural historical values.

The laboratory revolution was neither a paradigm shift nor a changing of the guard. It was, rather, a period of slow transition marked by the blending of traditions. When Vinogradskii decided to study botany with Famintsyn and not Beketov, he signaled his preference for the newly popular plant physiology over the seemingly staid morphological approach. For Vinogradskii's generation, physiology elicited visions of the imagination the heroic efforts of Robert Mayer, Claude Bernard, and

L. Ackert, *Sergei Vinogradskii and the Cycle of Life: From the Thermodynamics of Life to Ecological Microbiology, 1850-1950*, Archimedes 34,
DOI 10.1007/978-94-007-5198-9_11, © Springer Science+Business Media Dordrecht 2013

Louis Pasteur to bring laboratory science to bear on the nature of life—to tease apart the delicate fabric of vital processes. By choosing plant physiology, Vinogradskii had chosen a science that brought together three of the most experimental biological sciences in the second half of the nineteenth century: physiology, biochemistry, and bacteriology.

Like his teachers Famintsyn, Beketov, and even Pasteur, Vinogradskii had grown up in a culture bathed in naturalistic holism. Vinogradskii came from the land. Growing up in a southern Russian *pomeshchik* family that earned its wealth from agriculture, sugar refining, and banking; and being musically trained, for Vinogradskii, cycles—seasonal, industrial, and economic, and harmonious—resonated especially strongly. Moreover, when he studied the natural sciences at St. Petersburg University, these colloquial and vague cyclical referents came alive as a scientific conceptualization of the cycle of life.

Step-by step, Vinogradskii found a way to express his natural historical vision of the cycle of life in the language of the laboratory. He introduced the analytic and observational power of the laboratory into the wild of nature: first as Famintsyn's apprentice, then in De Bary's laboratory, and later in Zurich. Under the ocular of his microscope, in the microbial landscapes of his Petri dishes and the elective cultures of his retorts, Vinogradskii achieved a synthesis of the natural and the experimental. At the turn of the twentieth century, for him, the synthesis came to completion in the study of microbial nutrients cycling in nature. Soil science and geobotany had assimilated his research, and his students were continuing his labors. Had he glided quietly out of science at that time, he would still have enjoyed the admiration of his peers and an important place in the history of science.

While he was living what he called "the life of a latent scientist," the approach Vinogradskii had developed over the 1880s–1890s lay dormant. Like the microbial spores he investigated, his approach awaited a time when new conditions would again breathe life into it. That time came when Vinogradskii resurfaced in the early 1920s. Haeckel's term "oecology" of 1866 was now much in vogue in the biological sciences. Ecology was emerging as a self-conscious discipline, celebrated by the publication of new journals, papers, and textbooks; the offering of new courses at universities; and the foundation of new scientific institutions. Although these efforts appear to have crystallized around the simple principle that ecology is the study of natural relationships, early ecologists struggled to define a coherent agenda. In this melee, Vinogradskii successfully promoted as an ecological method his cycle of life perspective and the methods he developed to investigate it experimentally.

In the 1920s, soil scientists, microbiologists, plant ecologists, and forestry scientists—whose investigations required them to deal with the complexity of the soil—sought ways to meet the obligations of ecological thinking. Raised within the tradition of Vinogradskii's microbiology, and familiar with, or themselves now investigating, the autotrophic bacteria he had discovered; they found themselves learning again from a living classic. Vinogradskii's ecological methods of the 1920s stood on the scientific authority he had earned in the 1890s, and now they reaffirmed and extended that authority. His work found and created new audiences—ecologists adopted his methods and new disciplines formed around his approach.

Using a biographical approach, I have brought together a constellation of historical questions that considers each from a new perspective. Constrained, however, by the scope of Vinogradskii's life, this work merely touches on questions that deserve fuller treatment. His life in science spanned six of the most explosive decades in Russian and Western European social and political history, and in the history of science. The broad social movements that forced Vinogradskii along a particular historical path also provided impetus to more general trends in the history of Russian science. Vinogradskii, for example, did not represent the "men of the sixties," who were born in the 1840s, became inspired by the radical writers of their generation Pisarev and Chernyshevskii, and who saw science as a tool for improving Russian society. Never achieving their revolution, these "men of the sixties" instead became a force of modernizing liberal activists who operated within the existing framework of the Tsarist government. Vinogradskii's generation represented a continuation of that era. As "men of the eighties," they came of age in the 1880s during the reactionary reign of Tsar Alexander III and the era of small deeds.

These generations valued science highly, yet they produced different kinds of science. To what extent did the "men of the sixties" pass on their ideals to their successor generation? How did the efforts of both generations play out in the succeeding decades? An important first step in answering this question would be to understand how the men of the sixties continued social reform efforts through the training of new scientists. Dokuchaev and Beketov, for example, passed on more to their students than an interest in soil science and dynamic morphology. In what way did their courses reinforce or challenge the ideals—scientific or social—that their students brought from home? How did they teach them to blend the experimental with the practical? Whatever the details of this story, it is clear that their students applied these "sixties" values when developing new scientific approaches, institutions, and disciplines.

The large number of soil scientists who spent their formative years in southern Russia's agricultural regions raises the question of the influence of specific social conditions on the development of particular scientific disciplines. Was soil science in Russia fundamentally different from soil science in other countries? The same foundations that provided the basis for ecological sciences such as soil science, geobotany, and microbiology set the stage for the agrobiological hegemony of Trofim Lysenko. Did his scientific-political coup succeed in part because, by appropriating a longstanding tradition in Russian science, it found allies in across disciplines and generations? Soviet scientists referenced Vinogradskii—as they did other founding fathers of Russian science—in order to "justify and legitimate their institutional, intellectual, and career ambitions."[1] It is unlikely however, that Vinogradskii's case followed the other well-known examples related to the Michurinist campaigns of the 1930s. He was, after all, still alive, living in émigré status in France, and the

[1] On the phenomenon of "founding fathers" see Nikolai Krementsov, *Stalinist Science* (Princeton: Princeton University Press, 1995), 221–225. A variety of Soviet scientists regularly cited Vinogradskii's work, both from the pre- and post-revolutionary periods.

author of anti-Lenin and anti-Bolshevik newspaper articles. This episode does not reflect the experiences of Russian geneticists and so its exploration would add another dimension to our history of Stalinist science.

The history of soil science intersects all spheres of human activity—cultural, political, and scientific—yet it suffers from a nearly total lack of a sophisticated scholarship.[2] As slight as it is, this work, however, makes clear that Russian and Soviet soil science strongly influenced the development of that science in other countries. The widespread use of Russian terms "chernozem" and "podsol" in English-speaking countries represents more than a simple terminological transfer—intimately associated with those terms are such concepts of soil formation as "horizons" and "zones." An examination of the conceptual consistencies and contradictions in theoretical and practical soil science would help explain the transfer of knowledge across international scientific networks. It has been seen, for example, that Vinogradskii's discovery of autotrophism awakened an interest in soil scientists and geobotanists for microbiology. Building upon the foundation forged previously by agrochemists such as Pavel Kostychev, Vinogradskii created an experimental method for understanding the role of microbes in soil formation. American soil scientists—many of them southern Russians—also drew on Vinogradskii's work, but it remains unclear how his methods migrated through the American scientific community.

The consideration of Vinogradskii's contributions to microbiology opens a new perspective on the history of ecology; one that demands further explanation. Usually portrayed as emerging from the synthesis of Humboldtian phytogeography with Darwinian evolutionary theory by botanists in the late nineteenth and early twentieth century, ecology is also historically rooted in microbial physiology. Although historians of ecology recognize that early plant ecologists identified their science closely with plant physiology, these historians mistakenly limit plant physiology to the study of complex plants and plant communities. Eugene Cittadino and Joel Hagen have pointed out, for example, that the American ecologists Henry Cowles and Frederic Clements, and the British ecologist Arthur Tansley, considered the physiological work of Eugenius Warming, Oscar Drude, and A. F. W. Schimper "crucial to their new science."[3] Cittadino and Hagan gave no indication that these founding figures in the history of ecology devoted much research to investigating microbial fungi and bacteria.[4] The evolution of Vinogradskii's career from plant physiologist to ecologist demonstrates that early ecologists could consider physiology much more than the study of the nutritional needs of individuals or the formation of communities; it could encompass the entirety of nature—the cycle of life.

The story of how Vinogradskii transformed the cycle of life into a laboratory-based ecological perspective tells only one dimension of that concept's history.

[2] Following the pattern of the early history of medicine and science, the few works on the history of soil science have been written by practitioners.

[3] Cittadino, *Nature as Laboratory*, 150; Hagan, *An Entangled Bank*, 24–26.

[4] On Warming's microbiological work, see Chap. 2.

In his recent *The Cycles of Life: Civilizations and the Biosphere*, Vaclav Smil—like many other ecological microbiologists—traced the origin of the concept of a cycle of life to Lucretius's poem *De Rerum Natura*.[5] In Lucretius's description of the interconnectedness of all living and nonliving materials, modern scientists saw their interpretation of nature's seemingly obvious cyclical behavior.

Vinogradskii's speech on the cycle of life concept in 1896 falls within this tradition. He relied, however, on more proximate reference points. His sentiments echoed those made 50 years earlier by the French physiologist Jean-Baptiste Dumas and the botanist Ferdinand Cohn. In 1845, Dumas described the cycle of life as the eternal cycling of matter from plants to animals and back. Missing from his system, though, were microbes. In 1852, Cohn made "nature's smallest living beings" the central part of his vision of the cycle of life. A logical continuation of this work would be to investigate the religious and philosophical origins of the cycle of life and the process by which it migrated into science. This would require a broadening of my discussion of the nineteenth century botanists, physiologists, and microbiologists who inscribed it into their laboratory techniques, and to explore more extensively the influence of this tradition on twentieth century ecological sciences. Especially significant would be Lavoisier's writings on fermentation and nutrition, which represent one of the first appearances of the cycle of life in scientific discourse.

As this thesis shows, Vinogradskii transferred his cycle of life tradition to a diverse set of scientific schools. Numerous bacteriologists, soil scientists, forestry specialists, and public health workers adopted the cycle of life worldview and its inscription in Vinogradskii's techniques. It would be especially interesting to study how the cycle of life concept migrated with new microbiological techniques into American forestry, soil science, physiological chemistry, medical bacteriology, and public health. At the Yale University, for example, William Henry Brewer organized the Yale School of Forestry and provided the institutional and theoretical foundation for ecological forestry; Russell H. Chittenden applied his bacteriological training to help make the Sheffield Scientific School the fountain of American physiological chemistry, Charles-E. A. Winslow and his student Isidore S. Falk were major forces in the American Public Health Association and many other public health organizations worldwide; and Leo F. Rettger applied his skills and experience in bacteriological research and professional organizing to problems in human nutrition. These scientists' efforts to apply ecological methods in their research represent a conduit between the developments explored in this thesis and twentieth century developments in American bacteriology.

The history of the concept of the cycle of life in the twentieth century extends far beyond Vinogradskii. Vladimir Vernadsky and Vinogradskii shared a similar background—both came from an agricultural region in southern Russia, attended St. Petersburg University in the 1870s–1880s, and grounded their scientific worldview in Russian holism. Where Vinogradskii trained few students and expressed his

---

[5] Vaclav Smil, "Epigraph" to *The Cycles of Life: Civilizations and the Biosphere* (New York: Scientific American Library, 1997), vii.

scientific creativity through careful, tedious laboratory research, Vernadsky—who had a much more gregarious and aggressive scientific style—developed numerous and extensive research programs. Where Vinogradskii transformed the cycle of life concept into an ecological method for soil science, Vernadsky constructed the science of biogeochemistry and—as part of a sweeping theoretical framework for the entirety of earth science—a new conception of the biosphere.

One of the founders of ecosystem ecology, G. Evelyn Hutchinson, took great interest in Soviet biogeochemistry. A zoologist and limnologist by training, Hutchinson adopted and developed Vernadsky's holistic vision and methods, eventually training the first generation of ecosystem ecologists. Hutchinson trained some of the most influential ecologists, including Raymond Lindemann, Lynn Margulis, and Thomas Lovejoy. Hutchinson's story also illuminates the relatively unexplored theme of Russian influences on American science. A step in this direction would be to explore the interactions between Hutchinson's school and Soviet biogeochemists and other ecologically-minded scientists.[6] This formed the basis of a network of contacts between ecological thinkers in the United States and the Soviet Union, a network that spoke a common language based on Vinogradskii's ideas and methods.

Cycles populated Vinogradskii's life from his boyhood on the black earth of a Russian farm to his intellectual ferment at St. Petersburg University. They accompanied him on during his travels through the exotic landscapes of Alpine swamps and as he viewed his carefully prepared microscope slides, then to his laboratory at Russia's premier medical research institution, his scientific forestry on a Kiev estate, and finally to his resurrection as an ecologist in France. Vinogradskii discovered cycles in the microbial nutrition, in sulphur springs, in the soil, and in the biosphere. In a 60-year career in microbiology, he transformed the grand cycle of life into narrowly-focused nutritional cycles of individual species, and back out again, into physiological taxonomies of genera, and beyond there to nutrient cycles of the biosphere. Much of his work enjoyed his colleagues' respect. Acknowledging in his 50 years of research a plodding devotion and, at times, even genius, they always recognized the language of cycles as their own. Here, familiarity bred contentment. Why? Jorge Borges expressed it best when he wrote that "it may be that universal history is the history of the different intonations given a handful of metaphors." The fabric of human culture, language, and thought are woven from metaphors as enduring and ubiquitous as the cycle of life.

---

[6] Vladimir Vernadsky's son George—who was an eminent Russian historian at Yale—introduced Hutchinson to his father's work, and to the ecological microbiologist G. F. Gauze and ecologist A. N. Formozov.

Bibliography

Manuscripts Collections Visited

Omelianskii, Vasilii Leonaidovich, Arkhiv Russkoi Akademii Nauk, St. Peterburgskii Filial, fond 327.
Otdela Obshchei Mikrobiologii, Tsentral'nyi Gosudarstvennyi Istoricheskii Archiv, St. Petersburg, fond 2282.
Sankt-Peterburgskaia Konservatoriia, Tsentral'nyi Gosudarstvennyi Istoricheskii Archiv, St. Petersburg, fond 361.
Vinogradskii, Sergei Nikolaevich, Arkhiv Russkoi Akademii Nauk, Moskovskii Filial, fond 1601.
Waksman, Selman, Collections of the Manuscript Division, Library of Congress.
Winogradsky, Serge, Service des Archives de l'Institut Pasteur, Paris, France.

General Sources

Acot, Pascal. 1988. *Histoire de l'Écologie*. Paris: Presses Universitaires de France.
Adams, Mark B. (ed.). 1994. *The evolution of Theodosius Dobzhansky: Essays on his life and thought in Russia and America*. Princeton: Princeton University Press.
Allen, Garland. 1981. *Life science in the twentieth century*. Cambridge: Cambridge University Press.
Anker, Peder. 2001. *Imperial ecology: Environmental order in the British Empire, 1895–1945*. Cambridge: Harvard University Press.
Bailes, Kendall E. 1978. *Technology and society under Lenin and Stalin: Origins of the Soviet Technical Intelligentsia, 1917–1941*. Princeton: Princeton University Press.
Bailes, Kendall E. 1990. *Science and Russian culture in an age of revolutions: V. I. Vernadsky and his scientific school, 1863–1945*. Bloomington: Indiana University Press.
Benfey, Otto Theodor. Julius Lothar Meyer. *Dictionary of Scientific Biography* 9: 347–353.
Benison, Saul. 1976, Winter. René Dubos and the capsular polysaccharide of pneumococcus: An oral history memoir. *Bulletin of the History of Medicine* 50(4): 459–477.
Bevilacqua, Fabio. 1990. Helmholtz's Ueber die Erhaltung der Kraft: The emergence of a theoretical physicist. In *Hermann von Helmholtz and the foundations of nineteenth-century science*, ed. Cahan David, 291–333. Berkeley: University of California Press.
Bramwell, Anna. 1989. *Ecology in the twentieth century: A history*. New Haven: Yale University Press.

L. Ackert, *Sergei Vinogradskii and the Cycle of Life: From the Thermodynamics of Life to Ecological Microbiology, 1850-1950*, Archimedes 34,
DOI 10.1007/978-94-007-5198-9,

Brefeld, Oscar. 1872. Mucor mucedo, Chaetocladium jones'ii, Piptocephalis freseniana. Zygomyceten. *Botanische Untersuchungen über Schimmelpilze,* Heft 1, 1–64. Leipzig: Verlag von Arthur Felix.
Browne, Janet. 1983. *The secular ark: Studies in the history of biogeography*. New Haven: Yale University Press.
Bulloch, William. 1938. *The history of bacteriology*. London: Oxford University Press.
Bunsen, Robert. 1874. *Zeitschrift für analytische Chemie*. Heidelberg: Carl Winters Universitätsbuchhandlung.
Bunsen, Robert. 1904a. Ueber das Vorkommen von Gyps und Schwefel in Braunkohlenablagerungen. In *Studien des Göttingischen Vereins Bergmännischer Freunde* Bd. IV, 359, ed. Wilhelm Ostwald, (Notizenblatt, Nr. 6); republished in *Gesammlte Handlungen von Robert Bunsen.* Vol. 1. Leipzig: Verlag von Wilhelm Engelmann.
Bunsen, Robert. 1904b. Anleitung zur Analyse der Aschen und Mineralwasser. In *Gesammelte Abhandlungen von Robert Bunsen,* ed. Wilhem Ostwald and Max Bodenstein, *Auftrage der Deutschen Bunsen-Gesellschaft für angewandte physikalishe Chemie*, Band 3, 500–555. Leipzig: Verlag von Wilhelm Engelmann.
Büsgen, Moritz. 1889, January 4. A review of. S. Winogradsky, Beiträge zur Morphologie und Physiologie der Bacterien. Heft I. *Zur Morphologie und Physiologie der Schwefelbacterien* (Leipzig: Arthur Felix, 1888). *Botanische Zeitung* 47(1): 14–16.
Caneva, Kenneth L. 1993. *Robert Mayer and the conservation of energy*. Princeton: Princeton University Press.
Chekhov, Anton Pavlovich. 1948. Leshii. In *Polnoe Sobranie Sochinenii i Pisem A.P. Chekhova*, ed. A. M. Egolina i N. S. Tikhonova, 368–437. Moskva: OGIZ, Tom XI, *P'esy, 1885–1904*.
Cittadino, Eugene. 1990. *Nature as the laboratory: Darwinian plant ecology in the German Empire, 1880–1900*. Cambridge: Cambridge University Press.
Cohn, Ferdinand. 1872. Untersuchungen über Bacterien. *Beiträge zur Biologie der Pflanzen* (Breslau: J.U. Kern's Verlag, Max Muller), 1(2): 127–222.
Cohn, Ferdinand. 1875. Untersuchungen über Bacterien II. Idem., 1(III): 141–204.
Cohn, Ferdinand. 1939. *Bacteria, the smallest living beings*. Baltimore: Johns Hopkins University Press.
Cohn, Pauline, and Felix Rosen. 1901. *Ferdinand Cohn: Blätter der Erinnerung*. Breslau: J.U. Kern's Verlag.
Coleman, William. 1971. *Biology in the nineteenth-century: Problems of form, function, and transformation*. New York: Wiley.
Coleman, William. 1986, Summer. Evolution in ecology? The strategy of warming's ecological plant ecology. *Journal of the History of Biology* 19(2): 181–196.
Comfort, Nathaniel. 2001. *The tangled field: Barbara McClintock's search for the patterns of genetic control*. Cambridge, MA: Harvard University Press.
Conn, Harold Joel. 1916, December 15. The relative importance of fungi and bacteria in the soil. *Science N.S* 44(1146): 857–858.
Conn, Harold Joel. 1917. The proof of microbiological agency in the chemical transformation of soil. *Science* 44(1185): 252–255.
Conn, Harold Joel. 1918, January. The microscopic study of the bacteria and fungi in soil. *New York Agricultural Experiment Station Technical Bulletin* 64: 1–20.
Cooper, Jill Elaine. 1998. Of microbes and men: A scientific biography of René Jules Dubos. Ph.D. diss., Rutgers University.
Coulter, John Merle. 1910. *A textbook of botany for colleges and universities*. New York: American Book Company.
De Bary, Anton. 1879. *Die Erscheinung der Symbiose*. Strassburg: K.J. Trübner.
Doetsch, Raymond N. 1960. *Microbiology: Historical contribution from 1776 to 1908 by Spallanzani, Schwann, Pasteur, Cohn, Tyndall, Koch, Lister, Schloesing, Burrill, Ehrlich, Winogradsky, Warington, Beijerinck, Smith, Orla-Jensen*. New Brunswick: Rutgers University Press.
Dokhman, G.I. 1973. *Istoriia Geobotaniki v Rossii*. Moskva: Izd-vo Nauka.
Dokuchaev, Vasilli Vasil'evich. 1952. *Russkii Chernozem*. Moskva: Gosudarstvennoe Isd-vo Sel'skokhoziaistvennoi Literatury, Vtoroe Izdanie.

Drude, Oscar. 1913. *Die Ökologie der Pflanzen*. Braunschweig: Druck und verlag von Friedr. Viewig und Sohn.

Dubos, René. 1954. The fate of microorganisms in vivo. In *The biochemical determinants of microbial disease*, ed. Rene J. Dubos. Dubos: Harvard University Press.

Dubos, René. 1990. Each a part of the whole. In *The world of Rene Dubos: A collection of his writings*, ed. Piel Gerard and Osborn Segerberg Jr.. New York: Henry Holt and Company.

Dumas, Jean-Baptiste. 1842. *Essai de Statique Chimique des Êtres Organisés*, leçon le 20 août 1841, 2nd éd., Paris: Fortin, Masson, and Ce, Librairies.

Ehtard, A., and L. Olivier. 1882. De la réduction des sulfates par les être vivants. *Comptes Rendus* 846–849.

Elkana, Yehuda. 1974. *The discovery of the conservation of energy*. London: Hutchinson Educational, Ltd.

Engelmann, Theodor Wihelm. 1888, October 19. Die Purpurbacterien und ihre Beziehungen zum Licht. *Botanische Zeitung*, 46(42): 661–669; No. 43: 677; No. 44: 693–701; No. 45: 709–720.

Famintsyn, A. S. 1860. Organismy na granitse zhivotnago i rastitel'nago tsarstva. In *Sbornik Izdavaemyi Studentami Imeratorskago Peterburgskago Universiteta*, Vyp. 2. St. Peterburg: Tipografiia II-go Otd. Sob. E.I.B. Kantseliarii.

Famintsyn, A. S. 1865. Deistvie sveta kerosinovoi lampy na Spirogyra orthospira Naeg. In *Deistvie Sveta na Vodorosli i Nekotorye Drugie Blizkie k nim Organismy,* 39–56. St. Petersburg.

Famintsyn, A. S. 1883. *Obmen Veshchestvo i Prevrashchenie Ehnergii v Rasteniiakh* [The exchange of matter and the transformation of energy in plants]. Sankt-Peterburg: Imperatorskoi Akademii Nauk.

Famintsyn, A. S. Die Wirkung des lichtes auf Wachsen der Keimenden Kresse. *Memoires de l'Academie Sciences St. Petersburg*, Ser. 7, Tom 8, No. 15, 1–19.

Farley, John. 1982. *Gametes and spores: Ideas about sexual reproduction, 1750–1914*. Baltimore: The Johns Hopkins University Press.

Fischer, Hugo. 1909. Besitzen wir eine brauchbare Methode der bakteriologischen Bodenuntersuchung. *Zentralblatt für Bakteriologie, Parasitenkunde und Infektionskrankheiten*, Abt. 2, No. 23: 144–159.

Fresenius. 1854. *Jahrbucher des vereins fur Naturkunde in Herzogthum Nassau*, Heft. XI.

Fribes, V. A., with Serge Winogradsky. 1895. Sur le rouissage de lin et son agent microbien. *Comptes Rendus de L'Académie des Sciences* CXXI: 742.

Fruton, Joseph S. 1972. *Molecules and life: Historical essays on the interplay of chemistry and biology*. New York: Wiley-Interscience, a Division of John Wiley and Sons, Inc.

Geison, Gerald L. 1995. *The private science of Louis Pasteur*. Princeton: Princeton University Press.

Geison, Gerald L. Ferdinand Julius Cohn. *Dictionary of Scientific Biography* III: 336–341.

Gillespie, Charles Coulston. 1960. *The edge of objectivity: An essay in the history of scientific ideas*. Princeton: Princeton University Press.

Glas, Eduard. 1979. *Chemistry and physiology in their historical and philosophical relations*. Delft: Delft University Press.

Golley, Frank Benjamin. 1993. *A history of the ecosystem concept in ecology*. New Haven: Yale University Press.

Grigoriev, A.A. (ed.). 1946. *V. V. Dokuchaev i Geografiia, 1846–1946*. Moskva: S.S.S.R.: Izd-vo Akademii Nauk.

Hagan, Joel B. 1992. *An entangled bank: The origins of ecosystem ecology*. New Brunswick: Rutgers University Press.

Hantzsch, Arthur. 1890. *Bemerkungen über sterochemisch isomere Stickstoff-verbindungen*. Berlin: A. W. Schade's Buchdr.

Harman, Peter M. 1982. *Energy, force, and matter: The conceptual development of nineteenth-century physics*. Cambridge: Cambridge University Press.

Helms, Douglas, Anne B.W. Effland, and Patricia J. Durana (eds.). 2002. *Profiles in the history of the U. S. Soil Survey*. Ames: Iowa State Press.

Henderson, Lawrence T. 1923. Life and services of Louis Pasteur. *Proceedings of the American Philosophical Society* 62: iii–xiii.

Hereaus, A. 1886. Sur les Bactéries des Eaux de Source et sur les Propriétés Oxydantes. *Zeitschrift für Hygiene* 1: 193–234.

Hevly, Bruce. 1996. The heroic science of Glacier motion. In *Science in the field, Osiris*, vol. 11, ed. Kuklick Henrika and Robert E. Kohler, 66–86. Chicago: University of Chicago Press.

Höffding, Harald. 1955. *A history of modern philosophy: A sketch of the history of philosophy from the close of the renaissance to our day.* Trans: from the original German by B. E. Meyer. New York: Dover Publications.

Holmes, Frederick Lawrence. 1974. *Claude Bernard and animal chemistry: The emergence of a scientist*. Cambridge, MA: Harvard University Press.

Holmes, Frederick Lawrence. 1991. *Hans Krebs*, vol. 1. New York: Oxford University Press.

Hoppe-Seyler, Felix. 1877. Vorwort. *Zeitschrift fur physiologische Chemie* 1.

Ihde, Aaron J. 1964. *The development of modern chemistry*. New York: Harper & Row.

Ivanov, I.V. 2003. *Istoriia otechestvennogo pochvovedeniia: razvitie idei, differentsiatsiia, institutsializatsia. Kniga pervaia, 1870–1947*. Moskva: Nauka.

Kir'ianov, G.F. 1966. *Vasilii Vasil'evich Dokuchaev, 1846–1903*. Moskva: Izd-vo Nauka.

Kluyver, Albert J. 1931. *The chemical activities of micro-organisms*. London: University of London Press.

Koch, Alfred. 1891, October 2. A review of "Serge Winogradsky, Recherches sur les Organismes de la Nitrification," *Botanische Zeitung* 49(40): 669–672, 680–685, 698–701.

Kohler, Robert E. 2002. *Landscapes and labscapes: Exploring the lab-field border in biology*. Chicago: The University of Chicago Press.

Kossovich, I.S. 1949. Kratkii ocherk rabot i vzgliadov P. A. Kostychev v oblasti pochvovedeniia i zemledeliia. In *Pochvy chernozemnoi oblasti Rossii: Ikh proiskhozhdenie, sostav i svoistva*, ed. P.A. Kostychev, 198–223. Moskva: Gosudarstvennoe Izd-vo Sel'sko-khoziaistvennoi Literatury.

Kostychev, P.A. 1951. In *Izbrannye trudy*, ed. I.V. Tiurina. Moskva: Izd-vo Akademii nauk SSSR.

Kox, A.J., and Daniel M. Siegel (eds.). 1995. *No truth except in the details: Essays in honor of Martin J. Klein*. Dordrecht: Kluwer Academic Publishers.

Krainskii, A. Die Actinomyceten und ihre Bedeutung in der Nature. Zentrallblatt *Zentralblatt für Bakteriologie, Parasitenkunde und Infektionskrankheiten* 41(6): 649–688.

Krementsov, Nikolai. 1997. *Stalinist science*. Princeton: Princeton University Press.

Kremer, Richard L. 1990. *The thermodynamics of life and experimental physiology, 1770–1880*. New York: Garland Publishing, Inc.

Kuhn, Thomas S. 1969. Energy conservation as an example of simultaneous discovery. In *Critical problems in the history of science*, ed. Clagett Marshall, 321–356. Madison: University of Wisconsin Press.

Kupferberg, Eric. 2001. The expertise of germs: Practice, language, and authority in American bacteriology, 1899–1924. Ph. D. diss., Harvard University.

Kuznetsov, S.I. 1974. *Razvitie Idei S.N. Vinogradskogo v Oblasti Ekologicheskoi Mikrobiologii*. Moskva: Nauka.

Lenoir, Timothy. 1982. *The strategy of life: Teleology and mechanics in nineteenth century German biology*. Dordrecht/Holland/Boston: D. Reidel Publishing Co.

Lepovitz, Helena Wadley. Pilgrims, patients, and painters: The formation of a tourist culture in Bavaria. *Historical Reflections* 18(1): 121–145.

Lincoln, W.Bruce. 1989. *Red victory: A history of the Russian Civil War*. New York: Simon & Schuster, Inc.

Lipman, J. G. 1904. The fixation of atmospheric nitrogen by bacteria. *Proceedings of the twentieth annual convention of the association of official agricultural chemists,* 146–162. Washington, DC: Government Printing Office.

Mackaman, Douglas Peter. 1993. The landscape of a Ville d'Eau: Public space and social practice at the Spas of France, 1850–1890. *Proceedings of the Annual Meeting of the Western Society for French History* 20: 281–291.

Manoilenko, K.V. 1981. Rol' A. S. Famintsyn v Razvitii Ehvoliutsionnoi Fiziologii Rastenii. In *Andrei Sergeevich Famintsyn: Zhizn i Nauchnaia Deiatel'nost*, ed. A.L. Kursanov et al., 131–149. Leningrad: Nauka.

Mawdsley, Evan. 1987. *The Russian Civil War*. Boston: Allen & Unwin.
Mazing, Iu A., T.V. Andriushevich, and Iu A. Golikov. 2002. *Rasskazy o velikom bakteriologe S.N. Vinogradskom*. St. Peterburg: Rostok.
Mazumdar, Pauline M.H. 1995. *Species and specificity: An interpretation of the history of immunology*. Cambridge: Cambridge University Press.
McIntosh, Robert P. 1985. *The background of ecology: Concept and theory*. Cambridge: Cambridge University Press.
Mendelsohn, John Andrew. 1996, June. Cultures of bacteriology: Formation and transformation of a science in France and Germany, 1870–1914. Ph. D. diss., Princeton University.
Meyer, Lothar. 1864. Chemische Untersuchungen der Thermen zu Landeck in der Grafschaft Glatz. *Journal fur Praktische Chemie*. Band 91, Heft 1: 1–14
Meyer-Ahrens, Konrad. 1867. *Die Heilquellen und Kurorte der Schweiz: und einiger der Schweiz zunächst angrenzenden Gegenden der Nachbarstaaten*, 2nd ed. Zurich: Drell, Füssli, and Co.
Miliukov, Paul. 1962. *Russian and its crisis*. London: Collier Books.
Mirzoian, Eh N. 1981. Evoliutsionno-biokhimicheskie Vzgliady A. S. Famintsyn v sviazi s ego Filosofskimi i Obshchebiologicheskimi Vozzreniiami. In *Andrei Sergeevich Famintsyn: Zhizn i Nauchnaia Deiatel'nos*, ed. A.L. Kursanov et al., 150–164. Leningrad: Nauka.
Moberg, Carol L., and Zanvil A. Cohn. 1991, May. René J. Dubos. *Scientific American*: 66–72.
Morgenthaler, Erwin. 2000. *Von der Ökonomie der Nature zur Ökologie: Die Entwicklung ökologischer Denkens und seiner sprachlichen Ausdrucksformem* in *Philologische Studien und Quellen*, Heft 160. Berlin: Erich Schmidt Verlag.
Müller, D. Johannes Eugenius Bülow Warming. *Dictionary of Scientific Biography* XIV: 181–182.
Nicolson, Malcolm. 1996. Humboldtian plant geography after Humboldt: The link to ecology. *British Journal for the History of Science* 9(102): 289–310, Part 3. Cambridge: Cambridge University Press.
Nyhart, Lynn K. 1995. *Biology takes form: Animal morphology and the German Universities, 1800–1900*. Chicago: The University of Chicago Press.
Omelianskii, Vasilii L. 1909. *Osnovy Mikrobiologii*. St. Peterburg: Tipografia Imperatorskoi Akademii Nauk.
Omelianskii, Vasilii L. 1927. Sergei Nikolaevich Vinogradskii: Po povodu 70-letiia so dnia rozhdeniia. *Archiv Biologicheskikh Nauk* 28: 11–33.
Partington, J.R. 1960. *A short history of chemistry*, 3rd ed. New York: Harper and Brothers.
Pasteur, Louis. 1863, Juin. Recherches sur la putréfaction. *Comptes Rendus de l'Académie des Sciences,* séance du 29, LVI: 1189–1194.
Pasteur, Louis. 1876. *Etudes sur la Bière, ses maladies, causes qui les provoquent, procédé pour la rendre inaltérable; avec une théorie nouvelle de la fermentation*. Paris: Gauthier-Villars.
Pasteur, Louis. 1939. Fermentations et générations dites Spontanées. In *Oeuvres de Pasteur*, vol. II, ed. Vallery-Radot Pasteur. Paris: Masson et Cie, Éditeurs.
Pfeffer, Wilhelm. 1897. *Pflanzenphysiologie: Ein Handbuch der Lehre vom Stoffwechsel und Kraftwechsel in der Pflanzen.* Leipzig: Verlag von Wilhelm Engelmann. Zweite Auflage, Erster Band.
Piel, Gerard, and Osborn Segerberg Jr. (eds.). 1990. *The world of Rene Dubos: A collection of his writings*. New York: Henry Holt and Company.
Polevoi, V.V., et al. 1981. A. S. Famintsyn i Fiziologia Rastenii v Peterburgskom-Leningradskom Universitete. In *Andrei Sergeevich Famintsyn: Zhizn i Nauchnaia Deiatel'nost'*, ed. A.L. Kursanov, 56–85. Leningrad: Nauka.
Purrington, Robert D. 1997. *Physics in the nineteenth century*. New Brunswick: Rutgers University Press.
Rabinbach, Anson. 1990. *The human motor: Energy, fatigue, and the origins of modernity*. Berkeley: University of California Press.
Razumovskaia, Z.G. 1960. Zhizn' i nauchnaia deiatel'nost' Vasiliia Leonaidovicha Omelianskogo. In *Russkie Mikrobiologi S.N. Vinogradskii i V. L. Omelianskii*, ed. G.L. Seliber, 17–25. Moskva: Izadatel'stvo Ministerstva Sel'skogo Xoziaistva SSSR.

Reikhberg, G.E. (ed.). 1931. *Bor'ba za Nauku v Tsarskoi Rossii*. Moskva/Leningrad: Gosudarstvennoe Sotsial'no-Ehkonomicheskoe Izdatel'stvo.
Resse, Max. 1870. *Botanische Untersuchungen über die Alkoholgährungspilze*. Leipzig: Verlag von Arthur Felix.
Robbins, Richard G. 1975. *Famine in Russia, 1891–1892: The Imperial Government Responds to a Crisis*. New York: Columbia University Press.
Robinson, Gloria. Heinrich Anton De Bary. *Dictionary of Scientific Biography* 1: 612–613.
Rocke, Alan J. 1984. *Chemical atomism in the nineteenth century: From Dalton to Cannizzaro*. Columbus: Ohio State University Press.
Romell, Lars-Gunnar. 1932. A nitrosocystis from American forest soil. *Svensk Botanisk Tidskrift* 26(1–2): 303–312.
Romell, Lars-Gunnar. 1936. Winogradsky's quest for a method in soil microbiology. *Zentralblatt für Bakteriologie, Parasitenkunde und Infektionskrankheiten*. Abt. II, 93: 442–448.
Romell, Lars-Gunnar. Luftväxlingen i marken som ekologisk faktor. *Meddelanden från Statens Skogsförsöksanstalt*. 19(1), Stockholm.
Rosen, George. 1959. The conservation of energy and the study of metabolism. In *The historical development of physiological thought*, ed. Mc Brooks Chandler and Paul F. Cranefield. New York: The Hafner Publishing Company.
Russell, Edward J. 1921. In *Soil conditions and plant growth*, The Rothamsted monographs on agricultural science, 4th ed, ed. Dr E.J. Russell. London: Longmans, Green and Co.
Russell, Edward J. 1923. *The microorganisms of the soil*. London: Longmans, Green, and Co.
Russell, Edward J. 1956. *The land called me; An autobiography*. London: George Allen & Unwin, Ltd.
Russell, Edward J. 1966. *A history of agricultural science in Great Briton, 1620–1954*. London: George Allen & Unwin, Ltd.
Sapp, Jan. 1990. The nine lives of Gregor Mendel. In *Experimental inquiries*, ed. H.E. Le Grand, 137–166. Dordrecht/Boston: Kluwer Academic Publishers.
Schimper, A.F.W. 1898. *Pflanzen-Geographie auf Physiologischer Grundlage*. Jena: Verlag von Gustav Fischer.
Schloesing, J., and A. Müntz. 1877. Sur la Nitrification par les Ferments Organisés. *Comptes Rendus de l'Academie des Sciences* 84(7): 301–303; 85(22): 1018–1020; 1878, 86(14): 892–895; 1879, 89(21): 891–894; (25): 1074–1077.
Schulze, Ernst. 1888. Sur les matériaux de réserve et plus particulièrement sur les tannin des feuilles persistantes. *Annales agronomiques* 14: 525.
Schulze, Ernst. 1895. Les engrais verts en sols pauvres. *Annales agronomiques* 21: 394–396.
Seliber, G.L. (ed.). 1960. *Russkie Mikrobiologi S.N. Vinogradskii i V.L. Omelianskii*. Moskva: Izadatel'stvo Ministerstva Sel'skogo Xoziaistva SSSR.
Senchenkova, E.M. Issledovaniia A.S. Famintsyna po Fotosyntezu. In *Andrei Sergeevich Famintsyn: His life and scientific activity*, ed. Kursanov et al., 86–109.
Shortland, Michael, and Richard Yeo (eds.). 1996. *Telling lives in science: Essays on scientific biography*. Cambridge: Cambridge University Press.
Sibirtsev, N.M. 1949. Pamiati P.A. Kostychev: K godovomu dniu konchiny Pavla Andreevicha Kostycheva. In *Pochvy chernozemnoi oblasti Rossii: Ikh proiskhozhdenie, sostav i svoistva*, ed. P.A. Kostychev. Moskva: Gosudarstvennoe Izd-vo Sel'sko-khoziaistvennoi Literatury.
Smil, Vaclav. 1997. *Cycle of life: Civilization and the biosphere*. New York: Scientific American Library.
Smit, Pieter. Albert Jan Kluyver. *Dictionary of Scientific Biography* 7: 405–407.
Soderqvist, Thomas. 2011. The seven sisters: Subgenres of bioi of contemporary life scientists. *Journal of the History of Biology* 44(4): 633–650.
Stocklöv, Joachim. 1988. *Arthur Rudolf Hantzsch im Briefwechsel mit Wilhelm Ostwald*, Vol. 21, *Berliner Beiträge zur Geschihte der Naturwissenschaften und der Technik*. Berlin: Ellen R. Swinne-Verl.
Strick, James. 2000. *Sparks of life: Darwinism and the Victorian debates over spontaneous generation*. Cambridge, MA: Harvard University Press.
Strogonov, B.P. 1996. *Andrei Sergeevich Famintsyn, 1835–1918*. Moskva: Nauka.

Sulloway, Frank J. 1997. *Born to Rebel: Birth order, family dynamics, creative lives*. New York: Vintage Books.
Teich, Mikulas. 1970. The foundations of modern biochemistry. In *The chemistry of life: Eight lectures on the history of biochemistry*, ed. Needham Joseph, 171–191. Cambridge: Cambridge University Press.
Theobald, David W. 1966. *The concept of energy*. London: E. and F. N. Spon, Ltd.
Todes, Daniel P. 1989. *Darwin without Malthus: The struggle for existence in Russian evolutionary thought*. New York: Oxford University Press.
Todes, Daniel P. 2001. *Pavlov's physiology factory*. Baltimore: The Johns Hopkins University Press.
Trass, Kh Kh. 1976. *Geobotanika: Istoriia i Sovremennye Tendentsii Razvitiia*. Leningrad: Izd.-vo Nauka.
Tschulok, S. 1910. *Das System der Biologie: Forschung und Lehre*. Jena: G. Fischer.
van Niel, C.B. 1995. The "Delft School" and the rise of general microbiology. In *Beijerinck and the Delft School of Microbiology*, xiii–xxvii. Delft: Delft University Press.
Vernadsky, Vladimir I. 1930. *Zhivoe Vechestvo*. Moskva: Gosizdat.
Vernadsky, Vladimir I. 1994. *Trudy po Geoximii.* A. A. Iaroshevskii, ed. Moskva: Nauka.
von Meyer, Ernst. 1898. *A history of chemistry from the earliest times to the present day, being also an introduction to the study of the science,* 2nd ed. Trans. George McGowan. London: Macmillan and Co., Ltd.
Voorhees, Edward B., and Jacob G. Lipman. 1907. A review of investigations in soil bacteriology, a U.S. Department of Agriculture. *Office of Experiment Stations Bulletin* 194: 6–7. Washington, DC: Government Printing Office.
Vucinich, Alexander. 1970. *Science in Russian culture, 1861–1917*. Stanford: Stanford University Press.
Waksman, Selman. 1922. Microorganisms concerned in the oxidation of sulfur in the soil. *Journal of Bacteriology,* 231–238, 239–256, 605–608, 609–616.
Waksman, Selman. 1932. *Principles of soil microbiology*, Secondth ed. Baltimore: The Williams & Wilkins Company.
Waksman, Selman. 1935, June–December. Jacob G. Lipman as an investigator: A chapter in the history of soil microbiology. *Soil Science* 40: 11–23.
Waksman, Selman. 1953. *Sergei N. Winogradsky: His life and work, the story of a great bacteriologist*. New Brunswick: Rutgers University Press.
Waksman, Selman. 1954. *My life with microbes*. New York: Simon & Schuster.
Waksman, Selman. 1966. *Jacob G. Lipman: Agricultural scientist and humanitarian*. New Brunswick: Rutgers University Press.
Waksman, Byron H. 2001, April. Oral interview with Lloyd Ackert.
Warming, Eugenius. 1876. Om Nogle ved Danmarks Kyster levende Bakterier. Kjöbenhavn: Bianco Lunos Bogtrykkeri. Aftryk af *Videnskabelige Meddelelser fra den naturhistoriske Forenin i Kjobenhavn,* 1875, No. 20–28.
Warming, Eugenius. 1895. *Plantamsfund: grundtrak of den ökologiska plantegeographi*. Copenhagen: Philipsens Forlag.
Warming, Eugenius. 1896. *Lehrbuch der Ökologischen Planzengeographie. Eine Einführung in die Kenntniss der Pflanyenvereine*. Trans. E. Knoblauch. Berlin: Borntraeger.
Weiner, Douglas R. 1988. *Models of nature: Ecology, conservation, and cultural revolution in Soviet Russia*. Bloomington: Indiana University Press.
Weiner, Douglas R. 1999. *A little corner of freedom: Russian nature protection from Stalin to Gorbachev*. Berkeley: University of California Press.
Winter, Georg. 1884. Die Pilze Deutschlands, Oesterreich und der Schweiz. In *Dr. L. Rabenhorst's Kryptogamen-Flora von Deutschland, Oesterreich und der Schweiz*. Leipzig: Verlag von Eduard Kummer. Zweite Auflage, Erster Band.
Zavarzin, G. A. 1989a. Sergei Nikolaevich Vinogradskii (1856–1952). Khemosintez: K 100-letniiu otkrytiia S. N. Vinogradskim, 5–21. Moskva: Nauka.
Zavarzin, G.A. 1989b. Sergei N. Winogradsky and the discovery of chemosynthesis. In *Autotrophic bacteria*, ed. Hans G. Schlegel and Botho Bowein, 17–32. Madison: Science Tech Publishers.

Zavarzin, G.A. 2010. *Tri zhizni velikogo mikrobiologa: dokumental'naia povest' o Sergee Nikolaeviche Vinogradskom*. Moskva: Librokom.
Zopf, Wilhelm. 1885. *Die Spaltpilze: Nach dem neuesten Standpunkte bearbeitet*. Breslau: Verlag von Eduard Trewendt.

Vinogradskii's Publications Cited

Vinogradskii, Sergei N. 1883. O vliianii vneshnikh uslovii na razvitie *Mycoderma vini*. In *Trudy Sankt-Peterburgskogo Obshchestva Estestvoispytatelei*, XVI, 2nd ser., 132–135.
Winogradsky, Sergius. 1887, August 5. Ueber Schwefelbacterien. *Botanische Zeitung* 31: 489–507.
Winogradsky, Sergius. 1888a. *Zur morphologie und Physiologie der Schwefelbacterien*. Leipzig: Arthur Felix.
Winogradsky, Serge. 1888b. Ueber Eisenbacterien. *Botanische Zeitung* 17: 261–270, see 261.
Winogradsky, Serge. 1889. Recherches physiologiques sur les sulphobacteries. *Annales de l'Institut Pasteur* 3: 49–60.
Winogradsky, Serge. 1890a. Sur les organismes de la nitrification. *Comptes Rendus de l'Academie des Sciences* 110: 1013–1016.
Winogradsky, Serge. 1890b, May 12. Sur les Organismes de la Nitrification. *Comptes Rendus des Seances de l'Académie des Sciences* 60(19): 1013–1016; Idem., 1891 63(2): 89–92.
Winogradsky, Serge. 1890c. Recherches sur les Organismes de la Nitrification. *Annales de l'Institut Pasteur* 4: 215–231; 257–275, 760–811; Idem., 1891. 5: 92–100, 577–616.
Winogradsky, Serge. 1893. Sur l'assimilation de l'azote gaseaux de l'atmosphère par les microbes, *Comptes Rendus de Séances de l'Académie Sciences*. 1er Semestre 116(24): 1385–1388; Idem., *Comptes Rendus hebdomadaires des Seances de l'Académie des Sciences* 118(7): 353–355.
Winogradsky, Serge. 1897. O Roli Mikrobov v Obshchem Krugovorote Zhizni. *Arkhiv Biologicheskix Nauk* 7(3): 1–27.
Winogradsky, Serge. 1922. Eisenbakterien als Anorgoxydanten. *Zentralblatt für Bakteriologie* 57.
Winogradsky, Serge. 1923, Juillet–Décembre. Sur la méthode directe dans l'étude microbiologique du sol. *Comptes Rendus Hebdomadières des Séances de l'Académie des Sciences*, Paris, Tome 167: 1001–1004.
Winogradsky, Serge. 1924a, April 7. Sur la Microflore Autochtone de la Terre Arable. *Comptes Rendus Hebdomadières des Séances de l'Académie des Sciences* 178(7).
Winogradsky, Serge. 1924b, February. Sur la méthode directe dans l'étude microscopique du sol. *Chemie et Industrie* 11(2).
Winogradsky, Serge. 1925, April. Études sur la Microbiologie du Sol, 1. Sur la Méthode. *Annales de l'Institut Pasteur* 39(4): 299–354.
Winogradsky, Serge. 1926, Juin. Études sur la Microbiologie du Sol: Sur Les Microbes Fixateurs D'Azote. *Annales de l'Institut Pasteur* 40(6): 455–520.
Winogradsky, Serge. 1927, Octobre. Revue Critique: Principes de Microbiologie du Sol. *Annales de l'Institut Pasteur* 41(10): 1126–1138.
Winogradsky, Serge, and en collaboration avec J. Ziemiecka. 1928. Études sur la Microbiologie du Sol: Sur le pouvoir fixateur des Terres. *Annales de l'Institute Pasteur* 42(1): 36–62.
Winogradsky, Serge. 1931. La Biologie du Sol. In *Le Mans*, 1–4. Paris: Imprimerie Monnoyer.
Winogradsky, Serge. 1932a, Janvier. Études sur la Microbiologie du Sol (cinquième mémoire): Analyse Microbiologique du Sol, Principes d'une Nouvelle Méthode. *Annales de l'Institut Pasteur* 48(1): 89–133.
Winogradsky, Serge. 1932b. L'état actuel du problème de la fixation de l'azote atmosphérique et ses récents progrès. *Ac. Agr. De France*, séance du 14 mai 1930, published in 1932.
Winogradsky, Serge. 1935, June–December. The method in soil microbiology as illustrated by studies on azotobacter and the nitrifying organisms. *Soil Science* 40: 59–76.

Winogradsky, Serge. 1938, Avril. Études sur la Microbiologie du Sol et Des Eaux: Sur la Morphologie et l'Écologie des Azotobacter. *Annales de l'Institut Pasteur* 60(4): 351–400.
Winogradsky, Serge. 1939. La Microbiologie OEcologique: ses principes et son précède. *Annales Agronomique* 41: 1–23.
Winogradsky, Serge. 1947. Principes de la Microbiologie Oecologique. *Antoine van Leeuwenhoek Journal of Microbiology and Serology*: Jubilee Volume issued in honor of Albert J. Kluyver, 12(1–4): 1–16.
Winogradsky, Serge. 1949. *Microbiologie du Sol: Problèmes et Méthodes, Cinquante ans de Recherches, Ouvres Complètes*. Paris: Masson et Cie.
Winogradsky, Serge. 1951. *Mikrobiologiia Pochvy: Problemy i Metody, Piat'desiat let issledovanii* [Microbiology of the soil: Problems and methods, fifty years of research], ed. A. A. Imshenetskii. Moskva: Akademiia Nauk.

Index

L. Ackert, *Sergei Vinogradskii and the Cycle of Life: From the Thermodynamics of Life to Ecological Microbiology, 1850-1950*, Archimedes 34,
DOI 10.1007/978-94-007-5198-9,